Lecture Notes in Computer Science 4638

Commenced Publication in 1973
Founding and Former Series Editors:
Gerhard Goos, Juris Hartmanis, and Jan van Leeuwen

Thomas Stützle   Mauro Birattari
Holger H. Hoos (Eds.)

Engineering Stochastic Local Search Algorithms

Designing, Implementing and Analyzing Effective Heuristics

International Workshop, SLS 2007
Brussels, Belgium, September 6-8, 2007
Proceedings

 Springer

Volume Editors

Thomas Stützle
Mauro Birattari
Université Libre de Bruxelles (ULB)
CoDE, IRIDIA
Av. F. Roosevelt 50, CP 194/6, 1050 Bruxelles, Belgium
E-mail: {stuetzle,mbiro}@ulb.ac.be

Holger H. Hoos
University of British Columbia
Computer Science Department
2366 Main Mall, Vancouver, BC, V6T 1Z4, Canada
E-mail: hoos@cs.ubc.ca

Library of Congress Control Number: 2007933306

CR Subject Classification (1998): E.2, E.5, D.2, F.2, H.2, I.1.2, I.2.8, I.7

LNCS Sublibrary: SL 1 – Theoretical Computer Science and General Issues

ISSN   0302-9743
ISBN-10   3-540-74445-2 Springer Berlin Heidelberg New York
ISBN-13   978-3-540-74445-0 Springer Berlin Heidelberg New York

Springer is a part of Springer Science+Business Media

springer.com

© Springer-Verlag Berlin Heidelberg 2007
Printed in Germany

Typesetting: Camera-ready by author, data conversion by Scientific Publishing Services, Chennai, India
Printed on acid-free paper   SPIN: 12111459   06/3180   5 4 3 2 1 0

Preface

Stochastic local search (SLS) algorithms enjoy great popularity as powerful and versatile tools for tackling computationally hard decision and optimization problems from many areas of computer science, operations research, and engineering. To a large degree, this popularity is based on the conceptual simplicity of many SLS methods and on their excellent performance on a wide gamut of problems, ranging from rather abstract problems of high academic interest to the very specific problems encountered in many real-world applications. SLS methods range from quite simple construction procedures and iterative improvement algorithms to more complex general-purpose schemes, also widely known as metaheuristics, such as ant colony optimization, evolutionary computation, iterated local search, memetic algorithms, simulated annealing, tabu search and variable neighborhood search.

Historically, the development of effective SLS algorithms has been guided to a large extent by experience and intuition, and overall resembled more an art than a science. However, in recent years it has become evident that at the core of this development task there is a highly complex engineering process, which combines various aspects of algorithm design with empirical analysis techniques and problem-specific background, and which relies heavily on knowledge from a number of disciplines and areas, including computer science, operations research, artificial intelligence, and statistics. This development process needs to be assisted by a sound methodology that addresses the issues arising in the various phases of algorithm design, implementation, tuning, and experimental evaluation. A similarly principled approach is key to understanding better which SLS techniques are best suited for particular problem types and to gaining further insights into the relationship between algorithm components, parameter settings, problem characteristics, and performance.

The aim of *SLS 2007, Engineering Stochastic Local Search Algorithms — Designing, Implementing and Analyzing Effective Heuristics* was to stress the importance of an integration of relevant aspects of SLS research into a more coherent engineering methodology and to provide a forum for research in this direction. The workshop brought together researchers working on various aspects of SLS algorithms, ranging from fundamental SLS methods and techniques to more applied work on specific problems or real-life applications. We hope that this event will lead to an increased awareness of the importance of the engineering aspects in the design and implementation of SLS algorithms, and that it will help to tie together existing activities and to seed new efforts in this promising research area.

The importance and the timeliness of the topic of SLS engineering is witnessed by the more than 50 submissions we received for this workshop. From these submissions, the 12 full and 9 short papers contained in this volume and presented at the workshop were chosen based on a highly selective and rigorous peer-reviewing process; each of them reports results of very promising, ongoing research efforts from, or highly related to, the budding area of SLS engineering. The workshop program was complemented by the Doctoral Symposium on Engineering Stochastic Local Search Algorithms, which was organized by Enda Ridge and Edward Curry, and five tutorials on important topics in SLS engineering given by well-known researchers in the field.

We gratefully acknowledge the contributions of everyone at IRIDIA who helped in organizing SLS 2007. Special thanks go to Enda Ridge and Edward Curry for their enthusiasm in organizing the doctoral symposium. We thank all researchers who submitted their work and thus provided the basis for the workshop program; the members of the Program Committee and the additional referees for their help with the paper selection process; the Université Libre de Bruxelles for providing the rooms and logistic support; and, more generally, all those who contributed to the organization of the workshop. Finally, we would like to thank COMP2SYS,[1] the Belgian National Fund for Scientific Research, and the French Community of Belgium for supporting the workshop.

June 2007

Thomas Stützle
Mauro Birattari
Holger H. Hoos

---

[1] A Marie Curie Early Stage Training Site funded by the European Commission; more information is available at http://iridia.ulb.ac.be/comp2sys.

Organization

SLS 2007 was organized by IRIDIA, CoDE, Université Libre de Bruxelles, Belgium.

Workshop Chairs

Thomas Stützle   Université Libre de Bruxelles, Belgium
Mauro Birattari   Université Libre de Bruxelles, Belgium
Holger H. Hoos   University of British Columbia, Canada

Program Committee

Thomas Bartz-Beielstein   Cologne University of Applied Sciences, Germany
Roberto Battiti   Università di Trento, Italy
Christian Blum   Universitat Politècnica de Catalunya, Spain
Marco Chiarandini   University of Southern Denmark, Denmark
Carlos Cotta   University of Málaga, Spain
Camil Demetrescu   Università La Sapienza, Italy
Luca Di Gaspero   Università degli Studi di Udine, Italy
Karl F. Doerner   Universität Wien, Austria
Marco Dorigo   Université Libre de Bruxelles, Belgium
Carlos M. Fonseca   University of Algarve, Portugal
Michel Gendreau   Université de Montréal, Canada
Jens Gottlieb   SAP AG, Germany
Walter J. Gutjahr   Universität Wien, Austria
Pierre Hansen   GERAD and HEC Montreal, Canada
Jin-Kao Hao   University of Angers, France
Richard F. Hartl   Universität Wien, Austria
Geir Hasle   SINTEF Applied Mathematics, Norway
David Johnson   AT&T Labs Research, USA
Joshua Knowles   University of Manchester, UK
Arne Løkketangen   Molde University College, Norway
Vittorio Maniezzo   Università di Bologna, Italy
Catherine C. McGeoch   Amherst College, USA
Daniel Merkle   Universität Leipzig, Germany
Peter Merz   Universität Kaiserslautern, Germany
Martin Middendorf   Universität Leipzig, Germany
Pablo Moscato   University of Newcastle, Australia
Luis Paquete   University of Algarve, Portugal
Steven Prestwich   University College Cork, Ireland

Günther Raidl   Vienna University of Technology, Austria
Celso Ribeiro   Pontifícia Univ. Católica do Rio de Janeiro, Brazil
Andrea Roli   Università degli Studi G. D'Annunzio, Italy
Jonathan Rowe   University of Birmingham, UK
Ruben Ruiz   Valencia University of Technology, Spain
Michael Sampels   Université Libre de Bruxelles, Belgium
Andrea Schaerf   Università degli Studi di Udine, Italy
El-Ghazali Talbi   University of Lille, France
Pascal Van Hentenryck   Brown University, USA
Stefan Voss   University of Hamburg, Germany
Jean-Paul Watson   Sandia National Labs, USA
Ingo Wegener   Universität Dortmund, Germany
David Woodruff   University of California, Davis, USA
Mutsunori Yagiura   Nagoya University, Japan

Local Arrangements

Prasanna Balaprakash   Université Libre de Bruxelles, Belgium
Carlotta Piscopo   Université Libre de Bruxelles, Belgium

Additional Referees

Prasanna Balaprakash
Frank Hutter
Marco A. Montes de Oca

Sponsoring Institutions

COMP2SYS, Marie Curie Early Stage Training Site
  http://iridia.ulb.ac.be/comp2sys

National Fund for Scientific Research, Belgium
  http://www.fnrs.be

French Community of Belgium (through the research project ANTS)
  http://www.cfwb.be

Table of Contents

Short Papers

The Importance of Being Careful

Arne Løkketangen

Molde University College, Molde, Norway
Arne.Lokketangen@himolde.no

Abstract. Metaheuristic (and other) methods are often implemented using standardized parameter values. This is most often seen when comparing a favorite method with competing methods, but also when inexperienced researchers implement a method for the first time. Often the (hidden) correlations between the search method components and parameters are neglected or ignored, using only standardized templates. This paper looks at some of these pitfalls or hidden correlations, using the mechanisms of tabu search (TS) as examples. The points discussed are illustrated by examples from the authors experience.

1 Introduction

When designing and implementing heuristics for discrete optimization problems, there are many choices to be made. These include search paradigms, search mechanisms, search parameters, test sets, etc. Often these choices are treated as independent of each other, even though most researchers acknowledge some interdependence between the different search mechanisms and associated parameter settings. Our experience is that the search mechanisms and parameters often interact, and at times in unforeseen ways. This is a difficult problem to disentangle, and often even to discover.

This paper attempts to highlight some of the interactions experienced by the author, while implementing the Tabu Search (TS) metaheuristic [3]. Others will have similar experiences. The reader should at least learn that caution is required when designing heuristics, and testing should be carried out to verify the hypothesis underlying each mechanism and parameter choice.

One early example is the publicity in the early days of TS by people claiming that the best value for Tabu Tenure (TT) was 7. This was mainly due to the (somewhat erroneous) linkage to the article *The Magical Number Seven, Plus or Minus Two: Some Limits on Our Capacity for Processing Information* by Miller [1], creating the impression that TS was governed by some of the same rules as human cognitive abilities. The claim of 7 being the best TT has long since been abandoned, noticing that good TT values are also correlated to instance size and neighborhood structure.

Many articles on how to design and test heuristics have been published, indicating that these processes are far from being well-defined. Two much-cited works are *How not to do it* by Gent and Walsh [2] and *Testing Heuristics: We have it all wrong* by Hooker [4]. These articles contains much good advice, and

T. Stützle, M. Birattari, and H.H. Hoos (Eds.): SLS 2007, LNCS 4638, pp. 1–15, 2007.
© Springer-Verlag Berlin Heidelberg 2007

the interested reader should acquaint themselves with them. From the article by Gent and Walsh, the following advice stands out: Make fast, repeatable code. That is, make the code fast, so that more tests can be done in the same amount of cpu-time, and make the code repeatable, as there is very often need to run tests anew, due to some error, or referee comment, or other. One of Hooker's main points is that we should forget much of the competition aspect, as this leads us to waste time on low-level engineering instead of inventing or discovering new mechanisms. The engineering part is hardly research, but is often required by referees who will only accept articles presenting new *best* results.

One example of unnecessary slow code too often seen is when the TT mechanisms are implemented as a (circular) list. This is cumbersome, has a time complexity of $|TT|$ and is thus very time consuming, especially as the TT can be implemented in constant time with very little overhead.

The examples used for illustrations in the following sections are only explained in sufficient detail for understanding the arguments. To get more detail and background, the authors are referred to the original papers.

This paper is organized as follows. In Section 2 we look at values for the *Tabu Tenure*, while the usefulness of *Aspiration Criteria* is illustrated in Section 3. The question on whether to search in infeasible space or not is addressed in Section 4. Section 5 treats the *Move Selection*, Section 6 discusses *Learning and Forgetting*, and Section 7 discusses some tradeoff issues when implementing search methods. Finally the conclusions are summarized in Section 8.

2   The Tabu Tenure

The value for the Tabu Tenure is often given little consideration, and often it seems like it is taken out of thin air, or by reference to another article. (Occasionally one might even see the 7 +/- 2 argument). There is some controversy as to wether some randomization in the value of TT should be used, or not, with current consensus leaning towards some randomization. There are clear indicators, however, that the actual value of TT required is linked to the other mechanisms used, especially move evaluation, diversification and learning.

It is often stated that the purpose of the tabu status is in stopping the reversal of recent moves. This is clearly only part of the picture. An important effect of the tabu status is to block off most of the search space. As an example, in a bit-flip neighborhood, and a TT of 20, then the search space is reduced by a factor of 2^{20}. It is easy to construct tabu-criteria that are very powerful. As a different example, consider an edge-exchange neighborhood, often employed in graph problems. If e.g. the instance size is $|N|$ (same as the number of nodes) and the total number of edges is $|N|^2$, then both of the following tabu criteria might seem natural if both edges involved in a move becomes tabu:

1. A move is tabu if both edges of the move are tabu.
2. A move is tabu if either edge of the move is tabu.

The first criterion is very weak, while the second is very strong. These considerations are very often not present when research is presented in articles, but has

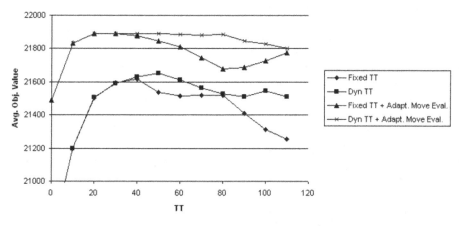

Fig. 1. Tabu Tenure

a large effect on the required values of the search parameters and the resulting search trajectories.

2.1 Varying Length TT

This subsection gives an example of when a dynamic TT is better than a fixed TT. The example is taken from Hvattum et al. [5]. The problem specification is given below to give the necessary background. The *Boolean Optimization Problem* can be regarded as a *Satisfiability Problem* with a profit on the variables. This profit is to be maximized.

$$max \qquad z = \sum_{j \in \mathbf{N}} c_j x_j \qquad (1)$$

$$s.t.$$

$$\sum_{j \in \mathbf{N}} a_{ij} x_j \geq 1, \qquad\qquad i \in \mathbf{M} \qquad (2)$$

$$x_j \in \{0, 1\}, \qquad\qquad j \in \mathbf{N} \qquad (3)$$

The TS metaheuristic developed in [5] to solve this problem uses a simple 1-bit flip neighborhood. In Figure 1 is shown the effect, using a single instance, on the search for various values of the TT, and keeping it either fixed or varying. For the varying TT, it was set to be a random number between 10 and the number on the x-axis. With a fixed move evaluation function, the results are as shown in the bottom pair of curves in the figure. As can be seen, having a dynamic TT is clearly beneficial, as better results are obtained over a larger range of values. The same conclusion is supported by the top two curves, giving the corresponding results using an adaptive move evaluation function (see 2.2). (Similar results were observed for the other test-cases).

2.2   Interaction of TT with Move Evaluation

In Figure 1 is also shown how the form of the move evaluation function influences both the TT and the overall search results. The search in [5] was designed to be able to search both in feasible and infeasible space. This is done by modifying the move evaluation function F_M with a component reflecting the resulting level of infeasibility. $F_M = \Delta V + w \cdot \Delta z$. The range of change in the objective function value component ΔV is scaled to ± 1, and the infeasibility component is the number of violated constraints multiplied by an adaptive weight, w. The two components are combined so as to give a balanced view to maintaining primal feasibility and a good objective function value. The emphasis between the two components is changed dynamically to keep the search around the feasibility boundary.

The value of w, the adaptive component, is initially set to 1. It is adjusted after each move as follows:

- If the current solution is feasible: $w = w + \Delta w_{inc}$
- If the current solution is not feasible, and $w > 1$: $w = w - \Delta w_{dec}$

As is illustrated in Figure 1, the effect of using an adaptive move evaluation function, capable of guiding the search into, and out of, infeasible space, has a large impact on the acceptable values for TT. The most important aspect of this is that the TT can be shorter, and good results are produced over a larger range of TT, making the search much more robust with respect to this parameter. (For the actual values used for Δw_{inc} and Δw_{dec}, see Section 4).

2.3   Localized Tabu Tenure

This subsection deals with the case when the global constraints can be regarded as weak. A typical case of this is described in *Interactive Planning for Sustainable Forest Harvesting* by Hasle et al. [6]. The object here is to device a sustainable plan for the treatment of each of a set of connected forest stands, or areas, in a 200 year perspective. Each stand has its own type of wood, age profile, soil quality, inclination, etc. giving rise to different growth rates and suitability for different types of trees. In Figure 2 is shown a typical area of forest, with around 700 stands. A treatment can typically be *cut* or *leave*, but there are also numerous other treatments like *thinning* and *gathering seeds*. There are global constraints of the type requiring yearly harvesting to be approximately equal each year (or follow a given profile). There are also neighboring constraints, like the *2-meter-constraint* that requires all neighboring stands to be at least 2 meters tall before harvesting of a given stand. A move is the changing of a treatment for a given stand, and the tabu criterion is not to make a move in a recently treated stand.

Preliminary testing of a TS heuristic for this problem showed that the best TT was nearly equal to the number of stands in the instance. This implies that most stands will have a tabu status at any given time, leaving very little room for the local search.

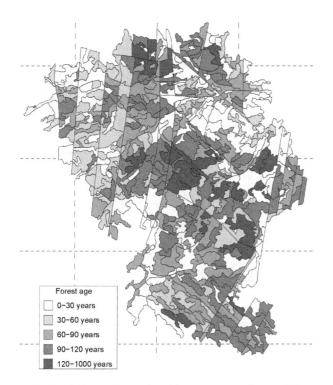

Forest age
☐ 0–30 years
30–60 years
60–90 years
90–120 years
120–1000 years

Fig. 2. Example of forest area and its subdivision into stands

The reason for this behavior is the weak nature of the global constraints, like balancing the harvesting over the years. Whenever the treatment plan for a given stand has been changed, the search moves to distant (in the geographical sense) stands. As there are many distant stands, all (or most) of them need to be tabu for the neighbor constraints to take effect, resulting in the extremely long required TT.

To remedy this, a device that might be called *Localized Tabu Tenure* can be employed. In this case the TT of a given stand is only counted for moves in the *vicinity* of the stand, where vicinity can be one or two stands distant. In this way, the TT will have a more natural length, and the neighboring constraints dealt with in the usual way.

3   Aspiration Criteria – Why Not Ignore It?

The inclusion of *Aspiration Criteria* (AC) in a TS is of crucial importance. The reason for this is that the basic TS mechanism of keeping certain attributes of the solution tabu for a certain time (i.e. iterations) often is too powerful, and blocks off attractive solutions. The mechanism called AS is designed to overcome this, by releasing the tabu status from moves leading to solutions having some predefined feature. The most common AC is to allow moves that leads to a new,

Table 1. The effect of using an Aspiration Criterion

Test case set	No ASP as % of w/ ASP	No Asp - Time to best	Asp - Time to best
Class 1	99.985	0.05	0.01
Class 2	100.000	0.01	0.00
Class 3	100.000	0.02	0.00
Class 4	100.000	0.02	0.00
Class 5	99.992	0.02	0.01
Class 6	100.000	0.02	0.00
Class 7	100.000	0.04	0.00
Class 8	99.985	0.03	0.03
Class 9	100.000	0.02	0.00
Class 10	100.000	0.03	0.00
Class 11	99.955	0.11	0.08
Class 12	99.980	0.06	0.01
Class 13	99.998	0.03	0.00

overall best, solution, even though the AC can take many forms. The effect of the AC is also dependent on the severeness of the Tabu Criterion.

AC are at times ignored, or only the standard AC outlined above is included. Very few reports the effect of their AC, and one might guess that in many cases the effect of the AC is not even tested. This might imply that the use of more sophisticated forms of AC including e.g. recency, segmentation, learning and forgetting is a largely untested field.

The benefit of AC will be illustrated by work from Hvattum et al. [5]. (This is the same as in Section 2.1).

In Table 1, is shown the effects of including an Aspiration Criterion. The AC chosen is the standard one of allowing a tabu move if it leads to a new *best* solution. Each row in the table contains the accumulated results for each of 15 classes of problems, ranging in size from 50 variables and 100 constraints to 1000 variables and 10.000 constraints. The three columns give the quality of the search when not using AC as a percentage of the results when including the AC, together with he time taken to reach the best solution with and without AC. As can be seen, the overall results are slightly better when using AC, and never worse. What is equally important is that time taken for the search to reach the best solution is much shorter when using the AC. This effect increases with problem size. Even if the quality is matched, the search takes longer.

4   Searching in Infeasible Space

One important design decision is whether the search should stay in feasible space all the time, or be allowed to traverse infeasible space on its hunt for good solutions. This decision of course depends on how easily feasibility can be recovered when going into infeasible space. For some problem types this is

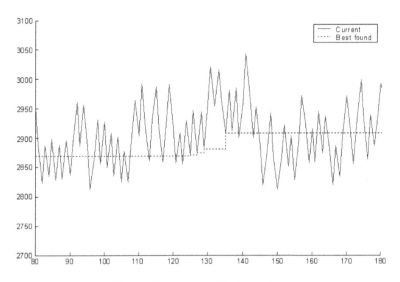

Fig. 3. Searching with dynamic w

trivial (e.g. 0/1 MCKP) , while for others it might be a major hurdle (e.g. ship scheduling).

One point to notice is that the optimal solution is always on the infeasibility boundary, so trying to let the search trajectory stay close to this might be a good thing. The feasible region might also be disconnected, as in Cappanera and Trubian [8].

The work by Hvattum et al. [5] is again used for illustrative purposes, this time for comparing feasible search with the same search being allowed to also search in infeasible space.

In Figure 3 is shown the TS as implemented in [5], on a small test-case for iterations 80 − 180. All the solution values above the dotted line showing the incumbent value are of course infeasible (as might some below the dotted line also be). The important thing to notice, is that whenever the incumbent is improved, the search comes from infeasible space.

This should be contrasted with the search depicted in Figure 4. This is using the same instance and TS settings as in Figure 3, but using a w value of 1. This restricts the search to stay mainly in feasible space. As is evident, the search is much more aggressive, and gets better results, when being allowed to wander into infeasible space. The proportion of the search time spent in infeasible space is determined by the values of the adaptive weight adjustment factors Δw_{inc} and Δw_{dec}, described in Section 2.2.

In Figure 5 are shown possible values for these two factors. As can be seen, it is the ratio of values that is important, not the absolute values. For the searches in [5] we chose $\Delta w_{inc} = 0.90$ and $\Delta w_{dec} = 0.35$. This behavior is also observed in other work.

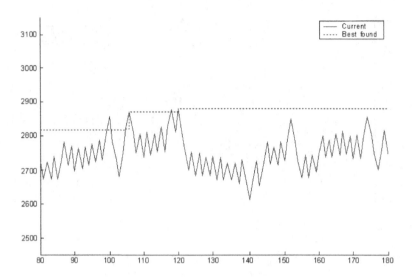

Fig. 4. Searching with static w

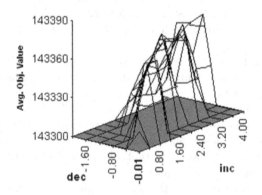

Fig. 5. Relationship between Δw_{inc} and Δw_{dec}

4.1   When Feasibility Is Difficult to Obtain

For some problems, even finding feasible solutions can be a hard task. One example of this is presented in *A Local Search Based Heuristic for the Demand Constrained Multidimensional Knapsack Problem* by Cappanera and Trubian [8]. This problem is based on a *0/1 MCKP*, but with additional cover constraints. It can also be called the *0/1 Multidimensional Knapsack/Covering Problem* (KCP) and formulated as follows:

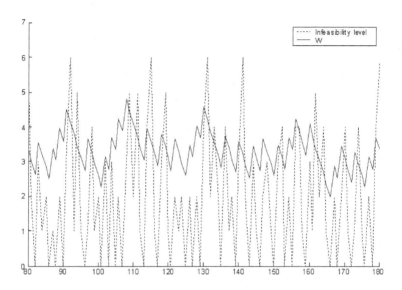

Fig. 6. Infeasibility levels and w

$$max \qquad z = \sum_{j \in \mathbf{N}} c_j x_j \qquad\qquad (4)$$

$$s.t.$$

$$\sum_{j \in \mathbf{N}} a_{ij} x_j \leq b_i, \qquad\qquad i \in \mathbf{M_A} \qquad (5)$$

$$\sum_{j \in \mathbf{N}} a_{ij} x_j \geq b_i, \qquad\qquad i \in \mathbf{M_B} \qquad (6)$$

$$x_j \in \{0, 1\}, \qquad\qquad j \in \mathbf{N} \qquad (7)$$

Here M_A and M_B represents the number of knapsack and cover constraints, respectively. The authors show that the set of feasible solutions might be disconnected with a bit-flip neighborhood. They employ a 2-phase method. In the first phase, all effort is in obtaining feasibility, ignoring the objective function value, while in the second phase, the search stays feasible all the time, using TS first with a single bit-flip neighborhood, and then a double bit-flip. They obtain good results, beating CPLEX on the larger problems, and finding more feasible solutions. The most difficult problems to find feasible solutions for, where those with the largest number of constraints.

The same problem was the topic for *Adaptive Memory Search for Multidemand Multidimensional Knapsack Problems* by Hvatum et al. [7]. Here the solution approach is very similar to the one used in [5], in that movement in infeasible space is allowed, but penalized. Computational experiments show that the new approach is in general able to find better solutions faster, and that feasible solutions now are found for all the problems in the test set.

This simple shift in design philosophy, going from separate phases in the first case (first feasibility, then objective function value), to an integrated approach in the second, illustrates the potential benefits of allowing infeasible moves when solving many problem types. This applies to both the case where feasibility is difficult to attain [7], and where obtaining feasibility is not an issue [5].

5   What Is the Best Move?

The normal move evaluation function is often very myopic, judging the quality of possible moves relative to the current solution. At the same time, the move selection function is usually designed to identify *good* moves in the neighborhood. This implies that the globally best move might not be at the top of the list, but will be among the top contenders. One way to exploit this, is by Probabilistic Move Acceptance, PMA, as described in Løkketangen and Glover [9].

This approach can be used by accepting randomly from the top of the list, but in a way biased towards the moves having the highest evaluations. The selection method is as follows: PMA:

1. Select a move acceptance probability, p.
2. Each iteration sort the admissible moves according to the move evaluation function.
3. Reject moves from the top of the list with probability $(1 - p)$ until a move is accepted.
4. Execute the selected move in the normal way.
5. If not finished, go to 2 and start the next iteration.

This can also be viewed as using randomness to diversify the search (as a substitute for deterministic use of memory structures), but in a guided way.

A different approach is taken in *Candidate List and Exploration Strategies for Solving 0/1 MIP Problems using a Pivot Neighborhood* by Løkketangen and Glover [10]. Here the idea is that since the neighborhood is expensive to explore (the neighborhood here is a LP-simplex pivoting neighborhood), several of the moves (pivots) may be good, and not just the one with the highest score. The problem here is that after a pivot is executed, the tableau is changed, and the other possible initial pivots might not be feasible any longer. In this case it turns out that, for our test-cases, up to 3 moves (or pivots) might beneficially be taken for the same initial tableau, as long as care is taken not to execute any move that has become illegal. The overall effect can thus be a substantial saving in time.

6   Learning and Forgetting

One of the guiding principles of Tabu Search is that the search should learn as it goes along. This is most often accomplished by using short-term memory

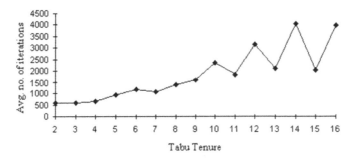

Fig. 7. Effect on Tabu Tenure of learning

for the tabu list, and frequency-based long-term memory for diversification (and intensification). Central to this is also the concept of *forgetting*. What is learnt is learnt in a certain context (i.e. area of the search space), and might not have much relevance later in the search. The short-term tabu list forgets quickly, but the long-term memory usually never forgets.

Løkketangen and Glover [11], [12] have implemented a TS for *Satisfiability Problems*, containing both constructive and iterative solvers, and various learning and forgetting aspects. The interested readers are referred to the papers for details.

This work contains a couple of points of interest for this paper. The first is the effect on the tabu tenure on forgetting. Learning was implemented by using extra clause weights indicating the importance of the individual clauses (constraints). These weights were updated every iteration by incrementing the weights for the violated clauses. *Forgetting* was implemented on top of this by multiplying the weights with a small discounting factor, slightly less than 1, thus diminishing the effect of the older learning. One very intersting side-effect of this is illustrated in Figure 7. The figure shows the required number of iterations to find a solution to a test problem for different values of tabu tenure. As is evident, the best results are when the TT is 2 or 3. One interpretation of this is that the TT is only required for blocking off the immediate move reversal, while the learning and forgetting seems enough to guide the search (of course in addition to the basic move evaluation function).

The second point of interest is illustrated in Figure 8. This figure illustrates the effect of learning between constructive runs, in that variables appearing in unsatisfied clauses (violated constraints) are given extra emphasis in the next constructive runs. An extra mechanisms that turned out to be important for the learning to work is the concept of *normalization*. This is, in essence, a mechanism that gives more importance to variables in shorter clauses, and the clauses shrink as varibles are assigned truth-values during the construction. Six different forms of normalization was tried and is shown in the figure on a simple test case. The severity of the importance is indicated by the numbers, with F1 being no normalization, and F6 being most severe.

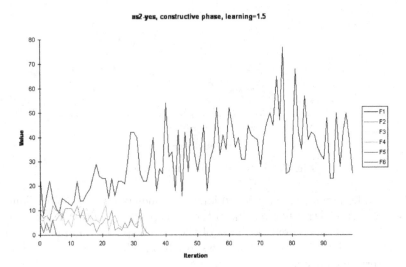

Fig. 8. Interaction between normalizations and Learning

As can be seen from the figure, the learning does not work without any normalization, and works better with the most severe type. This type of correlation is difficult to think of and to identify when testing heuristics, but should be kept in mind. Maybe what the search needs is an extra little twist that changes focus somewhat, as in this example.

7   Other Tradeoff Issues

It is evident that adding extra mechanisms increases the time needed per basic iteration of a search. Typical examples are long term diversification mechanisms added to a TS, or many of the examples mentioned in this paper. The trade-off is basically to either

- Do a lot every iteration, for a few iterations.
- Do a little every iteration, for many iterations.

A typical example of the first is TS, while Simulated Annealing (SA) is typical for the second. The design aim for the more complex searches must be that the added memory structures and mechanisms lets the search focus more rapidly on good solutions, offsetting the longer time per iteration with needing fewer iterations. (The same applies of course also to population-based methods, where global search space information is represented by a set of solutions.

This also touches on the topic of search time versus solution quality. It is clear that some real-world applications needs answers within milliseconds, while other applications may run for a long time, possibly overnight, or longer, before an answer is needed. Some searches need a larger set of iterations to trigger

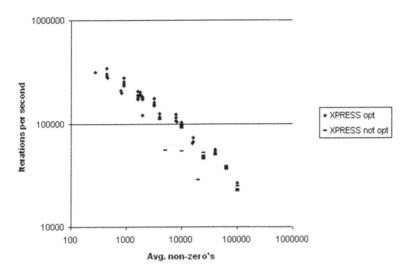

Fig. 9. Tradeoff between instance size and total iterations done

long term mechanisms, while others give good answers quickly, but with a lower potential of improvement.

Artificially generated instances from the academic world suffers from a lack of maximum time allowable, so only the *best* results are of interest, while second-best results seem to be admissible only if they are obtained *much* faster. Here it is about time to cite Hooker [4] again, and remind ourselves that we should not spend too much effort on the low-level engineering details, but rather focus on finding out when and where to apply our (new) mechanisms. (It is difficult to compete with the man-years of engineering effort put into CPLEX).

Computational testing is often done on a set of test-cases, some being small, some being big (these sizes increase rapidly over time as both hardware and methods get better and faster). The problem here is how much time should be allotted to the big instances compared to the small.

Ideally we want a method where search time per iteration increases linearly with size. Big instances require disproportionally more time, as both the time per iteration increases, and the number of iterations needed increases also. Often the small instances can be run as often as we like, taking virtually no time, while the bigger instances can only be run a few times.

Different authors treat his problem differently. Some allot the same amount of time or iterations to all sizes, while others use some (more or less artificial) argument to allot proportionally more time to the large instances. As an example, in Figure 9 is plotted the number of iterations in the 5 seconds alotted the TS heuristic of Hvattum et al. [5]. These 5 seconds where allotted all the searches. (They also had a series of tests giving 60 seconds to each search, but that only supports the argument.) The graph also shows on which instances Xpress-MP found the optimum within 4 hours. The horizontal axis gives the number of non-zero's in the problem instance (this measure takes into account both the number of variables and constraints), while the vertical axis then gives the corresponding number of iterations. As can be seen, the smallest problems managed to do around 300.000 iterations per second, while the largest only gets 1/1000 of this. This is in a sense turning things upside down, as we should rather give the large instance 1000 times more search effort. It is also evident from the computational results that better solutions might be expected with increased search time. (As things were at the time, the results beat CPLEX, Xpress and the rest of the competitors, so the results were sufficient).

8   Conclusions

This paper illustrates the importance of being careful and thorough when designing search heuristics. In particular the reader should focus on possible interactions between the different search mechanisms, and their associated parameter values. The examples used are from TS settings, but designers of other major metaheuristics must have similar experiences.

Also of importance are the somewhat novel techniques used for illustrative purposes, in the sense that if a mechanism does give better results, then it should be used, even though normal conventions are violated (as is the case in [10]).

References

1. Miller, G.A.: The Magical Number Seven, Plus or Minus Two: Some Limits on Our Capacity for Processing Information. The Psychological Review 63, 81–97 (1956)
2. Gent, I.P., Grant, S.A., MacIntyre, E., Prosser, P., Smith, B.M., Walsh, T.: How Not To Do It. Research Report 97.27, School of Computer Studies, University of Leeds (1997)
3. Glover, F., Laguna, M.: Tabu Search. Kluwer Academic Publishers, Dordrecht (1997)
4. Hooker, J.N.: Testing Heuristics: We have it all wrong. Journal of Heuristics 1(1), 33–42 (1995)
5. Hvattum, L.M., Løkketangen, A., Glover, F.: Adaptative Memory Search for Boolean Optimization Problems. Discrete Applied Mathematics 142, 99–109 (2004)
6. Hasle, G., Haavardtun, L.J., Kloster, O., Løkketangen, A.: Interactive Planning for Sustainable Forest Harvesting. Annals of Operations Research 95, 19–40 (2000)
7. Arntzen, H., Hvattum, L.M., Løkketangen, A.: Adaptive Memory Search for Multidemand Multidimensional Knapsack Problems. Computers & Operations Research 33, 2508–2525 (2006)
8. Cappanera, P., Trubian, M.: A Local Search Based Heuristic for the Demand Constrained Multidimensional Knapsack Problem. INFORMS Journal on Computing 17(1), 82–98 (2005)
9. Løkketangen, A., Glover, F.: Solving zero-one mixed integer programming problems using tabu search. European Journal of Operational Research 106, 624–658 (1998)

10. Løkketangen, A., Glover, F.: Candidate List and Exploration Strategies for Solving 0/1 MIP Problems using a Pivot Neighborhood. In: Voss, S., Martello, S., Osman, I.H., Roucairol, C. (eds.) Metaheuristics: Advances and Trends in Local Search Paradigms for Optimization, pp. 141–155. Kluwer Academic Publishers, Dordrecht (1999)
11. Løkketangen, A., Glover, F.: Surrogate Constraint Analysis - New Heuristics and Learning Schemes for Satisfiability Problems. In: Satisfiability Problem: Theory and Applications. DIMACS Series in Discrete Mathematics and Theoretical Computer Science, vol. 35 (1997)
12. Løkketangen, A., Glover, F.: Adaptive Memory Search Guidance for Satisfiability Problems. In: Rego, C., Alidae, B. (eds.) Metaheuristic Optimization Via Adaptive Memory and Evolution: Tabu Search and Scatter Search, pp. 213–227. Kluwer Academic Publishers, Dordrecht (2005)
13. Løkketangen, A., Glover, F.: Probabilistic Move Selection in Tabu Search for 0/1 Mixed Integer Programming Problems. In: Osman, I.H., Kelly, J.P. (eds.) Metaheuristics: Theory and Applications, Kluwer Academic Publishers, Dordrecht (1996)
14. Løkketangen, A., Glover, F.: Tabu Search for Zero/One Mixed Integer Programming with Advanced Level Strategies and Learning. International Journal of Operations and Quantitative Management 1(2), 89–109 (1995)

Designing and Tuning SLS Through Animation and Graphics: An Extended Walk-Through

Steven Halim and Roland H.C. Yap

School of Computing, National University of Singapore, Singapore
{stevenha,ryap}@comp.nus.edu.sg

Abstract. Stochastic Local Search (SLS) is quite effective for a variety of Combinatorial (Optimization) Problems. However, the performance of SLS depends on several factors and getting it right is not trivial. In practice, SLS may have to be carefully designed and tuned to give good results. Often this is done in an ad-hoc fashion. One approach to this issue is to use a tuning algorithm for finding good parameter settings to a black-box SLS algorithm. Another approach is white-box which takes advantage of the human in the process. In this paper, we show how visualization using a generic visual tool can be effective for a white-box approach to get the right SLS behavior on the fitness landscape of the problem instances at hand. We illustrate this by means of an extended walk-through on the Quadratic Assignment Problem. At the same time, we present the human-centric tool which has been developed.

1 Introduction

Stochastic Local Search (SLS) algorithms have been used to attack various NP-hard Combinatorial (Optimization) Problems (COP), often with satisfactory results. However, practitioners usually encounter difficulties when they try to get good performance from an SLS implementation.

Often the COPs encountered in industry are new COPs or variants of the classical COPs where no or little research is available. Sometimes a COP C' resembles a *classical* COP C for which a good SLS algorithm S is known. However, a direct implementation of S for C' usually will not immediately yield good performance. As there is no universal SLS algorithm which has good performance on all COPs, it is necessary to adapt existing SLS algorithm or create a new one for the COP at hand. It is often said that it is easy to create a working SLS for a COP, but hard to tune the SLS to achieve good performance on various instances of the COP [1,2,3,4,5]. [4] has summarized this situation as — 10% of the development time is spent in designing, implementing, and testing the SLS while the remaining 90% is used for (fine) tuning the SLS.

In several papers, SLS tuning is focused on tuning the parameter values. This, however, does not tell the whole story. From the beginning, one would already have a large number of combinations of SLS components (e.g. the neighborhood, etc) and search strategies (e.g. intensification, diversification, etc) to configure the SLS algorithm. Here we consider the SLS *design and tuning* problem as a

T. Stützle, M. Birattari, and H.H. Hoos (Eds.): SLS 2007, LNCS 4638, pp. 16–30, 2007.

holistic problem of finding a suitable configuration including parameter values, choice of components, and search strategies for an SLS algorithm in order to give good results for the class of COP instances at hand within *limited development time* and *limited running time*. This definition takes into account the practical efforts and resource requirements of industrial problems.

In [5], we classified the existing works on SLS design and tuning problem into two major approaches. Both approaches require the interaction of the human (the SLS algorithm designer) with the computer but in different ways:

1. The *black-box approach* favors the utilization of the computer for fine-tuning. The black-box approach aims to develop special 'tuning algorithms' to explore the SLS configuration space initially given by the algorithm designer. The best found configuration in the *given* configuration space is then returned. Examples of black-box approaches include F-Race [2], CALIBRA [4], +CARPS [6], Search Parameter Optimization [7], etc.
2. The *white-box approach* favors the utilization of human intelligence and/or visual perception strength. It aims to create tools or methods to help analyzing the performance of the SLS algorithm so that the algorithm designer has a basis for tweaking the SLS implementation. In this fashion, the SLS may be redesigned to extend beyond the initial configuration space. Examples of white-box approaches include Statistical Analysis (Fitness Distance Correlation Analysis, Run Time Distribution, etc) [3,8], Sequential Parameter Optimization [9], Human Guided Tabu Search [10], Visualization of Search Process [11,12], V-MDF and Viz [5,13,14,15], etc.

In this paper, we present a white-box approach where the algorithm designer first 'opens the white-box' using intuitive visualization tools. The objective is to investigate the *fitness landscape* [3,8,16] of the COP instance at hand and the SLS *trajectory* on the fitness landscape. This allows the algorithm designer to *understand* the kinds of problems which might be affecting the SLS, e.g. geting stuck in a deep local optimum of a rugged fitness landscape, failure to navigate to good regions, etc. The algorithm designer can use that knowledge to *roughly tune* the SLS either by using *known* or *new* heuristic tweaks. This process may involve not only changing parameter values but also the SLS algorithm design.

We show how visualization can be effective for designing and tuning an SLS algorithm in a white-box fashion. We have chosen as a starting baseline, a Robust Tabu Search (Ro-TS) algorithm which has been shown to give good performance for the Quadratic Assignment Problem (QAP) [17]. We show how by using *generic* visualization tools as an integral part of the development process, one can get insights into the fitness landscapes and problem characteristics of the instances of a COP. In turn, the insights inspire the development of an improved SLS that is suitable to navigate the fitness landscape.

Note: Several aspects of the presentation in this paper are best viewed via the animated video at this URL: `http://www.comp.nus.edu.sg/~stevenha/viz`. Here we have tried to illustrate the main points of our approach under the constraints imposed by a static, black and white paper (the online version of this paper in the Springer website is in color and can be magnified).

2   The Visualization Tool: Viz

We developed a visualization tool for the white-box approach called Viz. Viz currently consists of: (a) the Viz Experiment Wizard (EW); and (b) the Viz Single Instance Multiple Runs Analyzer (SIMRA).

The initial idea of using anchor points to explain SLS trajectory was first proposed in [5,13]. The proposal of a generic abstract 2-D visualization of the COP's fitness landscape using anchor points and visualization of SLS trajectory on it was proposed in the preliminary version of the Viz system [14]. The GUI and animation aspects of the visualizations in Viz are detailed in [15]. We remark that the Viz system is still undergoing development. As such, the visualizations and the GUI in this paper are significantly improved over [14,15].

Basic Concepts. Given a COP instance π, the *fitness landscape* of π, $FL(\pi)$, is defined[1] as $FL(\pi) : (S(\pi), d(s_1, s_2), g(\pi))$ [8], where $S(\pi)$ is the set of solutions of the COP instance (the search space), $d(s_1, s_2) : S \times S \rightarrow \Re$ is a distance metric which is natural for the COP in question (e.g. Hamming distance for QAP, see also the discussion in [18]), and $g(\pi) : S \rightarrow \Re$ is the objective function. One can visualize the abstract 2-D fitness landscape as a surface where each solution is a point on the surface, separated from other points according to its distance to those other points, and with a height that reflects the fitness (objective value) of that solution.

The *search trajectory* of an SLS algorithm on this $FL(\pi)$ is defined as a finite sequence $(s_0, s_1, \ldots, s_k) \in S(\pi)$ of search positions corresponding to solutions in $S(\pi)$ (a solution can be satisfiable or not satisfiable) such that $\forall i \in \{1, 2, \ldots, k\}$, (s_{i-1}, s_i) is a local move done by the SLS algorithm according to its neighborhood $N(\pi)$ [3]. The SLS algorithm can be visualized as a heuristic algorithm that navigates or walks through the fitness landscape of the COP instance in order to find higher (lower for minimizing COP) peaks (valleys) of the fitness landscape.

A Generic Fitness Landscape and Search Trajectory Visualization. [14,15] proposed that observing search trajectory on a fitness landscape is helpful to understand what is happening in the search. We describe the generic Fitness Landscape and Search Trajectory (FLST) visualization in Viz below:

First, we introduce the concept of Anchor Point (AP) set. The fitness landscape can contain an exponential number of solutions, thus we need an AP set which represents selected reference points on the fitness landscape to form an *approximate* fitness landscape visualization. The points along the SLS trajectory are then replayed back according to their distance w.r.t APs in the AP set.

In overview, FLST visualization is currently prepared using the following steps:

1. Potential AP Collection Phase (see Fig. 1)
In order to use FLST visualization in Viz, an SLS algorithm is augmented to save a history of the solutions visited into a log (i.e. RunLog file). In an SLS

---

[1] The definition that we use here [8] depends on the distance metric $d(s_1, s_2)$ and not the neighborhood $N(\pi)$ as in [3]. Thus, changing the $N(\pi)$ of the SLS will not change the fitness landscape but will change the SLS trajectory on that fitness landscape.

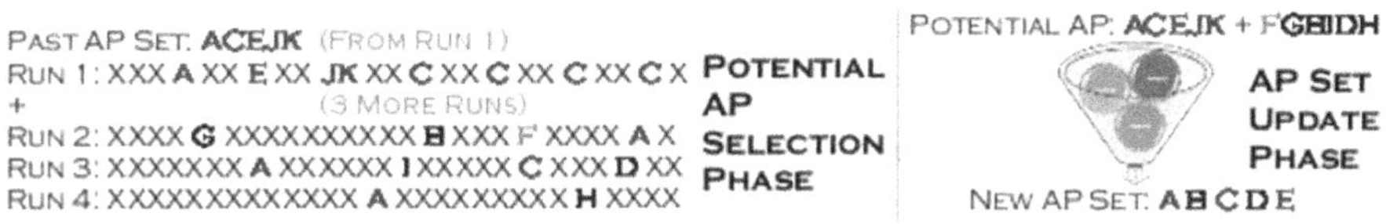

PAST AP SET: **ACEJK** (FROM RUN 1)
RUN 1: XXX **A** XX **E** XX **JK** XX **C** XX **C** XX **C** XX **C** X   **POTENTIAL**
+   (3 MORE RUNS)   **AP**
RUN 2: XXXX **G** XXXXXXXXX **B** XXX F XXXX **A** X   **SELECTION**
RUN 3: XXXXXXX **A** XX XXXX **I** XXXX **C** XXX **D** XX   **PHASE**
RUN 4: XXXXXXXXXXXXX **A** XXXXXXXX **H** XXXX

POTENTIAL AP: **ACEJK** + F**GBIDH**

AP SET
UPDATE
PHASE

NEW AP SET: **ABCDE**

Fig. 1. Potential AP Collection Phase and AP Set Update Phase. Let ACEJK be 5 APs collected from Run 1. Now, we execute 3 more SLS runs. The letters in red, green, blue indicate potential APs. After the AP Set Update Phase, 'B/D' replaces 'J/K'.

trajectory, the solution just before a non-improving move occurs is likely to be a local optimum solution. We use this feature to collect locally optimal solutions from the RunLog files from various SLS runs. These local optima solutions form a potential AP set.

2. AP Set Update Phase (see Fig. 1)
For visualization, it is not feasible to display too many points on screen. Thus, the AP set size should be limited to a small fraction of the possible solution set. It is important to only select meaningful APs that have these criteria:

1. Diverse: if AP X is included, then a similar AP X' will not.
2. High quality: good local optima must be included if found.
3. Important (best found, often visited) solutions from each run are included.

To achieve the above-mentioned criteria, we update the AP set according to the strategy below. The quality of the APs in the AP set is improved over time while the diversity of the APs is roughly maintained as we do more SLS runs:

1. Load past AP set from an AP log file if it exists.
2. If the AP set is still not full, *randomly* add any potential AP into it until it is full. The random addition of potential AP encourages diversity.
3. If the AP set is already full, compare each potential AP P with all current APs in AP set. Pick the nearest AP Q in term of distance metric w.r.t P. Then, if P is better than Q, P will replace Q. Otherwise, ignore P.
4. Save the current AP set into an AP log file for future experiments.

3. AP Layout Phase (see Fig. 2)
Now that we have a set of n-dimensional APs, how to visualize it? Our strategy is to use an abstract 2-D fitness landscape visualization which is guided by the distance metric between any two AP points. We use a spring model layout algorithm – which is similar to a statistical multi-dimensional scaling technique. Any two APs are connected by a virtual spring with its natural length equal to the distance between the two APs. The spring model will layout the points such that points that are near in the abstract 2-D fitness landscape are *approximately* near in the actual n-dimensional space and vice versa. The current implementation uses the NEATO algorithm from Graphviz [19] to do the AP layout.

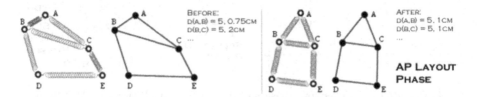

Fig. 2. AP Layout Phase using a spring model. This follows gestalt law of proximity [20], APs that are near (far) with each other are viewed as clustered (spread).

Fig. 3. AP Labeling Phase. This follows gestalt law of similarity [20]. APs with similar color and shape are grouped together by the human perception system. The perception laws of proximity and similarity are utilized in this fitness landscape visualization.

4. AP Labeling Phase (see Fig. 3)
Next, we label the APs based on their quality using the AP legend in Fig. 3. We use both color and shape to label the quality of the APs because even though color alone is easily seen on the screen, it is hard to see in black and white print. Newly added APs from recent runs are highlighted with a pink border so that the positions and qualities of the new APs can be easily identified.

The purpose of AP layout and labeling is to identify patterns in the fitness landscape at a glance, e.g. the distribution (clustered, spread, etc) and the quality (low variance ≈ smooth, high variance ≈ rugged, etc) of the old and new anchor points. This visualization extends the well known FDC scatter plot as it provides more information at a glance. It does not only measure distance and fitness w.r.t best found solution but *within all* APs in the AP set.

In Fig 7, we show an alternative side view of the abstract 2-D fitness landscape that emphasizes the smoothness or ruggedness of the APs.

5. Search Trajectory Layout Phase (see Fig 4)
Finally, we measure the distance between each point of the SLS run w.r.t to all points in the selected AP set. When the current point in the SLS run is near (by the distance metric) to one or more AP (this is possible since the points are n-dimensional), we draw a circle on those APs indicating that the current point is near those APs. Otherwise, we draw an indicator (see the example of a radar-like box in Fig 10, left) to tell how far is the current point to the nearest AP. This *nearness factor* is adjustable, ranges from 0 (exact match) to n (total mismatch), and is visualized via the radius of the enclosing circle. The drawing of the search trajectory as a *trail* is done over time using animation to indicate how the search progresses.

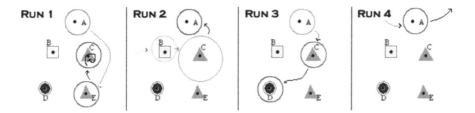

Fig. 4. Search Trajectory Layout Phase. The animation uses gestalt law of closure [20]. When the search trajectory is near an AP A and then quickly moves to AP B, human perception will fill in the details that the search traverses points along AP A→AP B. **Run 1**: SLS starts from a very bad AP A, walks to a medium quality AP E, and then cycles near AP C. **Run 2**: SLS starts from a bad AP B, then walks to a point near AP C (assume in Fig 1, the green point F in SLS run 2 happens to be near AP C), then moves to a very bad AP A (poor intensification). **Run 3**: SLS starts from a very bad AP A, but gradually walks to a medium AP C, escapes from it, then arrives at a good AP D (good intensification, compare this to Run 1 which is stuck in an attractive AP C). **Run 4**: The search trajectory of the SLS only bypasses a very bad AP A and is not near any other known APs (failure to navigate to promising region).

In Fig 4, we describe some possible interpretations of the search trajectories which are based on the four runs in Fig 1 and the fitness landscape in Fig 2 and Fig 3. As Fig 4 is static, we have represented the animation with arrows.

Remarks. Good visualization can help explaining information better than textual output alone. We note that for particular COPs, e.g. Traveling Salesman Problem (TSP), one could come up with a natural problem-specific visualizations, e.g. TSP tour visualization. However such visualizations may not show the features of the fitness landscape or search trajectory well and furthermore, some problems may not have a natural problem-specific visualization.

Viz Experiment Wizard (EW). To assist the user in preparing the SLS experiment design, executing the SLS runs, showing gross information about the runs, managing log files, and most importantly: to compute FLST visualization information as described above, we developed a tool called Viz EW; see Fig. 5.

Viz Single Instance Multiple Runs Analyzer (SIMRA). To visualize the fitness landscape of a COP instance and one (or more — currently only two) SLS trajectory on the same instance, we use Viz **S**ingle (COP) **I**nstance **M**ultiple (SLS) **R**uns **A**nalyzer (SIMRA). It uses the visualization information produced by Viz EW and animates it in real time like a VCR according to the search time recorded in the RunLog file. The visuals in Fig. 6, label 'A', give linked animations of FLST visualization as described in Fig. 1-4, objective value over time, FDC visualizations, algorithm specific visualizations (eg. tabu tenure over time), and problem specific visualizations (eg. QAP data matrices structure). It also displays textual information – see Fig. 6, label 'B'.

For other details about Viz GUI and visual features that are not included in this paper, please see http://www.comp.nus.edu.sg/~stevenha/viz or [5,13,14,15].

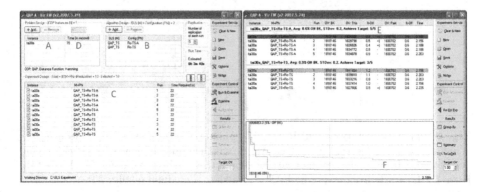

Fig. 5. Left: Viz EW has a GUI to specify: problem (**A**), algorithm (**B**), experiment (**C**) design, and time limit per instance (**D**). **Right**: Results can be grouped with *context-sensitive* statistical data in the group heading (**E**). The objective value summary visualization (**F**) gives a rough overview of the performance of the runs at a glance. The runs can be analyzed in detail with Viz SIMRA (see Fig. 6).

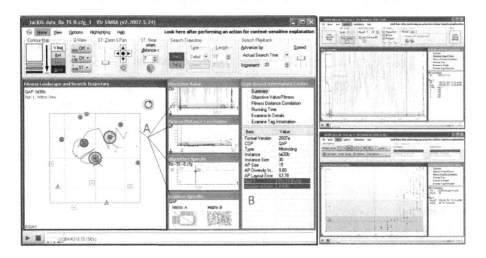

Fig. 6. The screen shot of Viz SIMRA for FLST and other visualizations

3   A Step by Step Walk-Through with Ro-TS for QAP

In this section, we give an *in-depth* walk-through which shows how to incorporate the visualizations in Viz into the entire design and tuning of the SLS development process. Our objective is to show how visualization allows us to more easily understand the COP fitness landscape and the SLS trajectory. This then allows new improvements on the insights gained from visualization. We also show Viz features for simplifying the SLS development process.

The COP used is the Quadratic Assignment Problem (QAP) with benchmark instances from QAPLIB [21]. We have purposely chosen a classic COP for this

Table 1. Initial Ro-TS-I Configuration

Component	Choice	Remark
Neighborhood	2-Opt	Natural (swap) move operation for QAP.
Objective Function	delta	Measure delta as shown in [17].
Tabu Tenure (TT)	n	The default tabu tenure length.
Tabu 'Table' [17]	pair $i - j$	Item i cannot be swapped with j for TT steps.
Aspiration Criteria	Better	Override tabu if move leads to better solution.
Search Strategy	'Ro-TS' [17]	Change TT within $[90\%*n\text{-}110\%*n]$ every $2n$ steps.

walk-through so that the reader can appreciate the walk-through and the results easier. We know the best known (BK) objective values for each QAP instances in this benchmark library. From this, we have defined the following solution quality measures: good ($< 1\%$-off BK), medium ($1\% - 2\%$-off BK), bad ($2\% - 3\%$-off BK), and very bad ($> 3\%$-off BK) points.

For the experiments, we have fixed the number of iterations of every SLS run to be quite 'small': $5n^2$ iterations, where n is the instance size. This is because state-of-the-art SLS algorithms (including the one we use) can obtain (near) optimal solutions for many QAPLIB instances with long runs. The goal of this experiment is to design and tune an SLS algorithm for attacking the selected QAP instances within the $5n^2$ iteration bound.

We remark that for the purpose of simplifying the walk-through and reducing the number of visuals, we have taken the liberty of presenting this walk-through from the final step viewpoint in order to fit the walk-through within page constraints. Most of the FLST visualizations make use of the final AP set from good and bad runs from the entire development process. In the actual development process, we learn the fitness landscape structure and the search trajectory behavior incrementally. In Section 3.7, we give an example of our learning process for obtaining the final AP set used in the walk-through.

3.1   Experiment Set-Up: QAP Instances and Baseline Algorithm

We pick Taillard's QAP instances: tai30a/30b/35a/35b/50a/50b as training instances and tai40a/40b/60a/60b as test instances. These artificial QAP instances are chosen because the real life instances in QAPLIB are quite 'small' (≤ 36). The selected instances vary in size and have been generated using two strategies: uniformly generated matrices proposed in 1991 [17] and non-uniform matrices which resemble real-life instances proposed in 1995 [22]. For our purposes, we should imagine that initially we are unaware of the distinction between these two types, perhaps we are only at the 1991 time point.

There are several (successful) SLS proposals for QAP in the literature, for example: Robust Tabu Search (Ro-TS) [17]. We use Ro-TS as our baseline SLS — it is already a good SLS for QAP. Our implementation, which we called Ro-TS-I, uses a configuration similar to Ro-TS. The details are in Table 1.

We conduct some pilot runs. Ro-TS-I results are given in Table 2. It seems that within the limited iteration bound, the initial results are reasonably good for

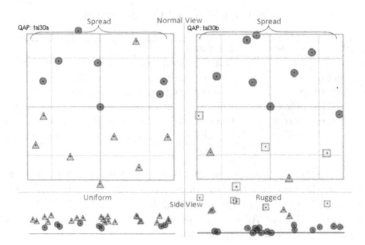

Fig. 7. Fitness landscape overview of two types of QAP instances (tai30a and tai30b) with the *same* label setting (see Fig. 3). The best found solution is always in the center. The anchor points in tai30a and tai30b are both spread throughout the fitness landscape and may be as far as the problem diameter (30 distance units). However, the quality of the anchor points in tai30b seems to be more 'rugged' than tai30a, especially when viewed using side view mode (the visualization is rotated 90 degrees along horizontal axis. The y-axis now shows the fitness of each AP w.r.t BK value).

tai30a/35a/35b/50a but not for tai30b/50b (underlined). We want to investigate why this happens and find out how to adjust Ro-TS-I.

3.2 Fitness Landscapes of QAP and Ro-TS-I Behavior

Using the final AP set (see Section 3.7) and with a Hamming distance metric [18], we observe the fitness landscape of QAP instances as shown in Fig. 7. We observe a significant difference between tai30a (and tai35a/50a) and tai30b (and tai35b/50b). See the figure text for details.

A working hypothesis from this observation is that there are at least two classes of QAP instances. The next question is how to identify which instance belongs to which class? Recall that we are not assuming that we know about the underlying problem instance generators. By looking at the fitness landscape visualizations and data matrices of the QAP instances, we classify tai30a/35a/50a (training), tai40a/60a (test) as QAP (type A) instances and tai30b/35b/50b (training), tai40b/60b (test) as QAP (type B) instances. This is consistent with the characteristics of these classes which differ in the smoothness (type A) or ruggedness (type B) in the fitness landscape visualization and in the uniformity (type A) and non-uniformity (type B) of their data matrices.

In Fig. 9 (left) we observe that Ro-TS-I already has reasonable performance on the QAP (type A) instances. This may be because the gap among local optima is small — the quality of most APs are either blue circle (good) or green triangle (medium). The animations of search trajectories do not indicate any obvious

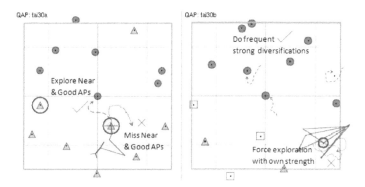

Fig. 8. Hypotheses of what the *idealized* good walks (blue dashed lines) should look like on QAP (type A/B) instances and the actual Ro-TS-I behavior (red dotted lines)

sign of Ro-TS-I being stuck in a local optimum (explained textually: the search trajectory is found to be near one AP, quickly escapes from it, and later appears to be near another *different* AP). However, none of the Ro-TS-I runs gets the best known objective value within the limited iteration bound.

However, the performance of the same Ro-TS-I on the QAP (type B) instances is very bad. In Fig. 10 (left) we observe that Ro-TS-I never visits any known good APs. Prior to obtaining the final AP set (see Section 3.7), we observe that Ro-TS-I often gets stuck in a bad local optimum region and cannot escape (explained textually: the search trajectory enters a region near one AP, then until the last iteration, it is still near the same AP). If that local optimum happens to be bad, the final best found solution reported will also be bad.

3.3   Hypotheses to Improve Walks on the QAP Fitness Landscapes

From the visualizations, the algorithm designers can arrive at the following two hypotheses on how the SLS should behave on these two classes:

(1). QAP (type A) landscape is more smooth, so it is hard to decide where to navigate as 'everything' looks good. Diversifying too much may not be effective since we will likely end up in another region with similar quality. Our hypothesis: it is better for the SLS to reduce the possibility of missing the best solution within a close region where the SLS is currently in. Fig. 8 (left) illustrates our hypothesis and shows the desired trajectory (blue dashed lines) searches around nearby good local optima rather than the trajectory of Ro-TS-I (red dotted lines) which moves away from the good local optima region.

(2). QAP (type B) landscape is more rugged, the local optima are deeper and spread out. Thus, we hypothesize that within the limited iteration bound, rather than 'struggling' to escape deep local optima with its own strength (e.g. via tabu mechanism), it is better for the SLS to do frequent strong diversifications as we know it is hard to escape from deep local optima and the 'nearest' local optima may be quite 'far' anyway. Fig. 8 (right) illustrates our hypothesis where the desired trajectory (blue dashed lines) only makes short runs in a region before

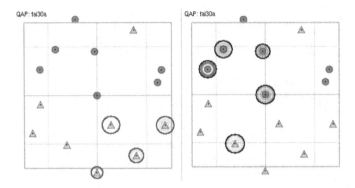

Fig. 9. Search coverage of Ro-TS-I (left) and Ro-TS-A (right) on QAP (type A) instance. Ro-TS-I (see large red circles) covers some of medium quality APs then wanders somewhere else far from known APs (animation not shown). On the other hand, lower Tabu Tenure Range in Ro-TS-A enables it to take different trajectory which covers some good quality APs (see large blue circles).

jumping elsewhere rather than the trajectory of Ro-TS-I (red dotted lines) which struggles to escape a deep local optimum.

3.4   Tweaking Ro-TS-I to Ro-TS-A for QAP (Type A) Instances

The search coverage (APs that are near any points in search trajectory are highlighted) of Ro-TS-I on QAP (type A) instances is not good (see Fig. 9, left).

Visualization shows that Ro-TS-I actually gets near to known good APs but does not in the end navigate to those APs. Thus the good APs are missed. One reason may be that during short Ro-TS-I runs ($5n^2$ iterations), there are some Ro-TS-I moves leading to such good APs that are under tabu status and are not overridden by aspiration criteria.

The idea for robustness in Ro-TS [17] is in changing the tabu tenure randomly during the search within a defined Tabu Tenure Range (TTR). The TTR is defined by Tabu Tenure Low (TTL) and Tabu Tenure Delta (TTD) as follows: TTR = [TTL, TTL + TTD]. To encourage more intensification, we modify the Ro-TS-I by decreasing TTR from the recommendation in [17], TTR = $[0.9n, 1.1n]$ into a lower range and changing the robust tabu tenure value more often (after n steps, not after $2n$ steps, see Table 2).

Since we do not have information on the best TTR for Ro-TS-I, except that it should be lower, we experimented with various TTR settings. TTR = $[0.4n, 0.8n]$ performed well on the training instances and also slightly better than the original Ro-TS-I (see Table 2).

We denote Ro-TS-I with lower TTR by Ro-TS-A. We observed that although TTR is smaller, it is still enough to ensure that Ro-TS-A avoids solution cycling. This may be because it is quite easy to escape from any local optima of smooth fitness landscape of QAP (type A) instances. Fig. 9 (right) shows the search coverage of Ro-TS-A which seems better than the search coverage of Ro-TS-I.

3.5   Tweaking Ro-TS-I to Ro-TS-B for QAP (Type B) Instances

Visualization reveals that Ro-TS-I does not manage to arrive near any of the known good APs. This leads to a very bad performance (see Fig. 10, left).

In Fig. 10 (left), if circle 'X' is outside circle 'Y', it means that no AP is near the current point of the search trajectory of Ro-TS-I. The inability of Ro-TS-I to arrive at any known good APs combined with our understanding of QAP (type B) fitness landscape (Section 3.7) tells us that Ro-TS-I is stuck in deep, poor quality local optima and fails to navigate to a good quality region.

We have learned that in order to make Ro-TS-I behave more like our hypothesis, we need to add a strong diversification strategy. We tweak Ro-TS-I such that after n non improving moves, Ro-TS-I is considered to be stuck in a deep local optimum. To escape from it, we employ a *strong* diversification mechanism which preserves $max(0, n\text{-}X)$ items and randomly permutates the assignment of the other $min(n, X)$ items in the current solution. The value of X must be sufficiently large: close to n but not equal to n, otherwise it would be tantamount to random restart. The rationale for this *strong* diversification heuristic is that we see in the fitness landscape that blue circle (good) APs in type B instances are located quite far apart but *not* as far as the problem diameter n.

Now, it remains to determine the value of the diversification strength X. We experimented with different values of X on tai30b and obtained good results when X=20. However, for other training instance tai50b, the best value for X is when X=30. Visualization shows that even with X=20 (out of diameter n=50), Ro-TS-I still has problem in escaping local optima of tai50b.

From this, we realize that X should not be *fixed* for *all* QAP (type B) instances but rather be *robust* within a range correlated with the instance size. After further experimentation, we arrived at a good range for X to be $[\frac{3}{4}n..\frac{7}{8}n]$, performing well on the training instances. The value of X will be randomly changed within this range after each diversification step.

We denote the revised SLS algorithm by Ro-TS-B. Animation shows that Ro-TS-B visits several APs with varying qualities, each for short time. Some APs visited by Ro-TS-B have good quality (see Fig. 10 (right) and Table 2).

3.6   Benchmarking on the Test Instances

Finally, we verify the results against the test instances (also with the same iteration bound). The results are given in Table 2. On average, Ro-TS-A performs slightly better than Ro-TS-I on QAP (type A) instances while Ro-TS-B is significantly better than Ro-TS-I on QAP (type B) instances. The results here are also comparable to the updated Ro-TS results in [22].

We see that applying either Ro-TS-A or Ro-TS-B to the other instance class has no or negative improvements (underlined). This shows that we have successfully tailored RoTS to match the different fitness landscapes of these instances.

3.7   The Learning Process

The fitness landscape of a COP instance is typically exponentially large. It is infeasible to enumerate all solutions just to analyze parts of the solutions visited

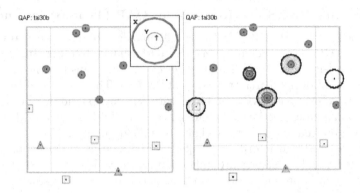

Fig. 10. Search coverage of Ro-TS-I (left) and Ro-TS-B (right) on QAP (type B) instance. Ro-TS-I (left) never arrives at any known good APs. On the other hand, Ro-TS-B (right) employs frequent strong diversifications and we see that it visits several APs of varying qualities (see large blue circles) that are (very) far from each other.

Table 2. The results of Ro-TS-I, Ro-TS-A, and Ro-TS-B on training and test instances – averaged over **5** runs per instance. The instance size n, Best Known (BK) objective value, maximum iteration bound $(5n^2)$, run times, average percentage deviation $\bar{x}$ from BK and its standard deviation σ are given.

Training Instances

Instance	n	Best Known	Iters	Time	Ro-TS-I $\bar{x}$	σ	Ro-TS-A $\bar{x}$	σ	Ro-TS-B $\bar{x}$	σ
tai30a	30	1818146	4500	3s	1.1	0.4	**1.0**	0.3		
tai35a	35	2422002	6125	5s	1.1	0.2	**1.1**	0.1		
tai50a	50	4938796	12500	18s	1.5	0.1	**1.4**	0.3		
tai30b	30	637117113	4500	3s	<u>16.1</u>	0.0			**0.2**	0.1
tai35b	35	283315445	6125	5s	1.8	0.0			**0.3**	0.0
tai50b	50	458821517	12500	18s	<u>7.5</u>	0.0			**0.7**	0.2

Test Instances

Instance	n	Best Known	Iters	Time	Ro-TS-I $\bar{x}$	σ	Ro-TS-A $\bar{x}$	σ	Ro-TS-B $\bar{x}$	σ
tai40a	40	3139370	8000	8s	1.4	0.1	**1.0**	0.5	<u>2.0</u>	0.2
tai60a	60	7205962	18000	34s	1.7	0.1	**1.6**	0.2	<u>2.2</u>	0.3
tai40b	40	637250948	8000	8s	9.0	0.0	<u>9.0</u>	0.1	**0.0**	0.1
tai60b	60	608215054	18000	35s	2.1	0.1	<u>2.9</u>	0.2	**0.3**	0.1

by the SLS trajectory. As with other white-box approaches, FLST visualization must rely on progressive learning based on the solutions found by non-exact approaches. Here, we show our learning example for QAP (type B) instances:

Starting with Ro-TS-I, we obtained a fitness landscape (see Fig. 11. **A**) with APs that are uniform, very bad (because we know the BK value), but not too far from each other (average pairwise distance between APs is $\approx \frac{1}{3}n$). We were curious why Ro-TS-I was stuck and added random restart diversification. Then, Ro-TS-I visited new (pink borders) and better APs (see Fig. 11. **B**) which are quite far from the previous very bad APs. These better points are located in

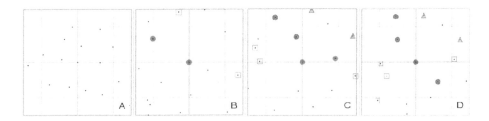

Fig. 11. The learning process for understanding QAP (type B) fitness landscape

a deep valley as the fitness gaps between them and their immediate neighbors are quite big. Soon, we learned that QAP (type B) has multiple deep valleys scattered far apart (see Fig. 11. **C, D**). The distance between good APs are quite far but not as much as n. This tells us that good APs have similar substructures which inspires the strategy in Section 3.5.

For instances where BK values are not known (especially for new problems), we know less but we can still evaluate SLS performance w.r.t to what we have, e.g. in QAP (type B) example above, finding a new (better) AP that is far from any poorer APs that are known so far indicates the need of diversification.

4   Conclusion

In this paper, we have described the Viz system and have shown an extended walk-through using Viz. The walk-through shows how to design and tune a baseline Ro-TS algorithm for the QAP. The insights from visualization results in two improved variants. We believe that the example is realistic and demonstrates why visualization is an interesting approach for designing and tuning SLS.

Viz is still under development but the prototype sytem can be downloaded from: http://www.comp.nus.edu.sg/~stevenha/viz.

Acknowledgements

We would like to thank the reviewers and the editors for their constructive inputs to improve the paper.

References

1. Charon, I., Hudry, O.: Mixing Different Components of Metaheuristics. In: Meta-Heuristics: Theory and Applications, pp. 589–603. Kluwer (1996)
2. Birattari, M.: The Problem of Tuning Metaheuristics as seen from a machine learning perspective. PhD thesis, Université Libre de Bruxelles (2004)
3. Hoos, H., Stützle, T.: Stochastic Local Search: Foundations and Applications. Morgan Kaufmann, San Francisco (2005)

4. Adenso-Diaz, B., Laguna, M.: Fine-tuning of Algorithms Using Fractional Experimental Designs and Local Search. Operations Research 54(1), 99–114 (2006)
5. Halim, S., Lau, H.: Tuning Tabu Search Strategies via Visual Diagnosis. In: Meta-Heuristics: Progress as Complex Systems Optimization, Kluwer (2007)
6. Monett-Diaz, D.: +CARPS: Configuration of Metaheuristics Based on Cooperative Agents. In: International Workshop on Hybrid Metaheuristics, pp. 115–125 (2004)
7. Hutter, H., Hamadi, Y., Hoos, H., Leyton-Brown, K.: Performance Prediction and Automated Tuning of Randomized and Parametic Algorithms. In: International Conference on Principles and Practice of Constraint Programming, pp. 213–228 (2006)
8. Merz, P.: Memetic Algorithms for Combinatorial Optimization: Fitness Landscapes & Effective Search Strategies. PhD thesis, University of Siegen, Germany (2000)
9. Bartz-Beielstein, T.: Experimental Research in Evolutionary Computation: The New Experimentalism. Springer, Heidelberg (2006)
10. Klau, G., Lesh, N., Marks, J., Mitzenmacher, M.: Human-Guided Tabu Search. In: National Conference on Artificial Intelligence (AAAI), pp. 41–47 (2002)
11. Syrjakow, M., Szczerbicka, H.: Java-based Animation of Probabilistic Search Algorithms. In: International Conference on Web-based Modeling and Simulation, pp. 182–187 (1999)
12. Kadluczka, M., Nelson, P., Tirpak, T.: N-to-2-Space Mapping for Visualization of Search Algorithm Performance. In: International Conference on Tools with Artificial Intelligence, pp. 508–513 (2004)
13. Lau, H., Wan, W., Halim, S.: Tuning Tabu Search Strategies via Visual Diagnosis. In: Metaheuristics International Conference, pp. 630–636 (2005)
14. Halim, S., Yap, R., Lau, H.: Visualization for Analyzing Trajectory-Based Metaheuristic Search Algorithms. In: European Conference on Artificial Intelligence, pp. 703–704 (2006)
15. Halim, S., Yap, R., Lau, H.: Viz: A Visual Analysis Suite for Explaining Local Search Behavior. In: User Interface Software and Technology, pp. 57–66 (2006)
16. Schneider, J., Kirkpatrick, S.: Stochastic Optimization. Springer, Heidelberg (2006)
17. Taillard, E.: Robust Tabu Search for the Quadratic Assignment Problem. Parallel Computing 17, 443–455 (1991)
18. Schiavinotto, T., Stützle, T.: A Review of Metrics on Permutations for Search Landscape Analysis. Computers and Operation Research 34(10), 3143–3153 (2007)
19. Graphviz: Graph Visualization Software, http://www.graphviz.org
20. Ware, C.: Information Visualization: Perception for Design. Morgan Kaufmann, San Francisco (2004)
21. QAPLIB: Quadratic assignment problem library, http://www.seas.upenn.edu/qaplib
22. Taillard, E.: Comparison of Iterative Searches for the Quadratic Assignment Problem. Location Science 3, 87–105 (1995)

Implementation Effort and Performance

A Comparison of Custom and Out-of-the-Box Metaheuristics on the Vehicle Routing Problem with Stochastic Demand

Paola Pellegrini[1] and Mauro Birattari[2]

[1] Department of Applied Mathematics, Università Ca' Foscari, Venice, Italy
[2] IRIDIA, CoDE, Université Libre de Bruxelles, Brussels, Belgium
paolap@pellegrini.it, mbiro@ulb.ac.be

Abstract. In practical applications, one can take advantage of meta-heuristics in different ways: To simplify, we can say that metaheuristics can be either used *out-of-the-box* or a *custom* version can be developed. The former way requires a rather low effort, and in general allows to obtain fairly good results. The latter implies a larger investment in the design, implementation, and fine-tuning, and can often produce state-of-the-art results.

Unfortunately, most of the research works proposing an empirical analysis of metaheuristics do not even try to quantify the development effort devoted to the algorithms under consideration. In other words, they do not make clear whether they considered *out-of-the-box* or *custom* implementations of the metaheuristics under analysis. The lack of this information seriously undermines the generality and utility of these works.

The aim of the paper is to stress that results obtained with *out-of-the-box* implementations cannot be always generalized to *custom* ones, and vice versa. As a case study, we focus on the vehicle routing problem with stochastic demand and on five among the most successful metaheuristics—namely, tabu search, simulated annealing, genetic algorithm, iterated local search, and ant colony optimization. We show that the relative performance of these algorithms strongly varies whether one considers *out-of-the-box* implementations or *custom* ones, in which the parameters are accurately fine-tuned.

1 Introduction

The term *metaheuristics* [1] recently became widely adopted for designating a class of approaches used for tackling optimization problems.

> *A metaheuristic is a set of algorithmic concepts that can be used to define heuristic methods applicable to a wide set of different problems.*
>
> [2, p. 25]

The generality of metaheuristics and the ease with which they can be applied to the most diverse combinatorial optimization problems is definitely the main reason for their success. A basic implementation of metaheuristic can be obtained

T. Stützle, M. Birattari, and H.H. Hoos (Eds.): SLS 2007, LNCS 4638, pp. 31–45, 2007.

with quite a low effort. Typically, with such an implementation, it is possible to achieve fairly good results. Nonetheless, it has been shown in the literature that metaheuristics can obtain state-of-the-art results with some larger implementation effort. To simplify, we can say that metaheuristics can be either used *out-of-the-box* or a *custom* version can be developed.

This flexibility of metaheuristics is definitely a positive feature: One can start with an *out-of-the-box* version of a metaheuristic for quickly having some preliminary results and for gaining a deeper understanding of the problem at hand. One can then move to a *custom* version for obtaining better performance without having to switch to a completely different technology.

Nonetheless, this flexibility has a downside: The fact that metaheuristics achieve different results according to the development effort can be reason of misunderstanding. In particular, it could well happen that, as we show in the case study proposed in this paper, a metaheuristic M_1 performs better than a metaheuristic M_2 on a given problem when *out-of-the-box* versions of M_1 and M_2 are considered; whereas M_2 performs better that M_1 on the very same problem when *custom* versions are concerned. In this sense, results obtained with *out-of-the-box* implementations do not always generalize to *custom* ones, and vice versa.

In the literature, this fact is often neglected and the development effort devoted to algorithms is rarely quantified. This can be partially justified by the fact that measuring development effort is not a simple and well-defined task. Nonetheless, without this piece of information, the usefulness of an empirical study is somehow impaired.

With this paper we wish to stress that two experimental studies, one performed in the *out-of-the-box* context, and the other in the *custom* one, may lead to different conclusions. To this aim, we consider as a case study the vehicle routing problem with stochastic demand, and five of the most successful metaheuristics—namely, tabu search, simulated annealing, genetic algorithm, iterated local search, and ant colony optimization. Our goal is to show that the relative performance of the above metaheuristics depends on the implementations considered.

In order to attenuate the problem concerning the different ability of a single designer in implementing various approaches, we consider the implementations of the five metaheuristics produced within the *Metaheuristics Network*,[1] a EU funded research project started in 2000 and accomplished in 2004. In the *Metaheuristics Network*, five academic groups and two companies, each specialized in the development and application of one or more of the above metaheuristics, joined their research efforts with the aim of gathering a deeper insight into the theory and practice of metaheuristics. For a detailed description of the metaheuristics developed by the *Metaheuristics Network* for the vehicle routing problem with stochastic demand, we refer the reader to [3].

In our analysis, these implementations are considered as *black-box* metaheuristics: By modifying their parameters, we obtain the *out-of-the-box* and the *custom* versions. The first ones are obtained by randomly drawing the parameters from

---

[1] http://www.metaheuristics.net/

a defined range. The second ones are obtained by fine-tuning the parameters through an automatic procedure based on the F-Race algorithm [4,5].

Selecting the best values for the parameters, given the class of instances that are to be tackled, is definitely a sort of customization. By using this element as the only difference between *custom* and *out-of-the-box* implementations, we are neglecting the customization of most of the components of metaheuristics. Nonetheless, the goal of the paper is to show that an analysis based on *custom* implementations might produce radically different results from one based on *out-of-the-box* implementations. If we succeed to show this fact when even one single element characterizing *custom* implementations is considered, namely the fine-tuning of parameters, we have nevertheless reached our goal.

It is interesting to note here that Bartz-Beielstein [6] has already addressed some of the issues discussed in the paper. In particular, he warns against performing a comparative analysis of algorithms that have not been properly fine-tuned. The results we present in the paper confirm the concerns raised in [6]. Nonetheless, our standpoint is somehow different from the one taken in [6]. Indeed, we think that analyzing the performance of *out-of-the-box* versions of metaheuristics can be in itself relevant since in some practical contexts the development of *custom* versions of a metaheuristic is not possible due to time and/or budget constraints.

The rest of the paper is organized as follows. In Section 2, we present a panoramic view of the literature concerning the vehicle routing problem with stochastic demand, the metaheuristics considered, and the tuning problem. In Section 3, we describe the specific characteristics of these elements as they appear in our analysis. In Section 4, the experimental study is reported. Finally, in Section 5, we make some conclusions.

2   Literature Overview

The three main topics of interest of our analysis are introduced in this section. We first focus on the vehicle routing problem with stochastic demand. Then we sketch the five metaheuristics considered, and the problem of fine-tuning metaheuristics.

The vehicle routing problem with stochastic demand (VRPSD) can be described as follows: Given a fleet of vehicles with finite capacity, a set of customers has to be served at minimum cost. The demand of each customer is *a priori* unknown and only its probability distribution is available. The actual demand is revealed only when the customer is reached. The objective of the VRPSD is the minimization of the total expected traveling cost.

Optimal methods, heuristics, and metaheuristics have been proposed in the literature for tackling this problem. In particular, the problem is first addressed by [7] in 1969. [8], [9] and [10] use techniques from stochastic programming to solve optimally small instances. [11] and [12] propose different heuristics for solving the VRPSD. They consider the construction of an *a priori* TSP-wise tour. This tour is then split according to precise rules. [13] propose a strategy for splitting the *a priori* tour allowing the restocking before a stockout, when this is profitable. Secomandi [14,15] analyzes different possibilities for applying dynamic programming to this problem. [16] and [17] tackle the VRPSD using metaheuristic approaches. In particular, [16] adopt simulated annealing while [17] use tabu search. Finally, an extended analysis on the behavior of different metaheuristics is proposed by [3]. Two classical local search algorithms have been used for the VRPSD: the Or-opt [13] and the 3-opt [3] procedures.

Following [3], we focus on five of the most popular metaheuristics: tabu search (TS), simulated annealing (SA), genetic algorithm (GA), iterated local search (ILS), and ant colony optimization (ACO).

Tabu search consists in the exploration of the solution space via a local search procedure. Non-improving moves are accepted, and a short term memory is used. The latter expedient is introduced in order to avoid sequences of moves that constantly repeat themselves [18].

Simulated annealing takes inspiration from the annealing process in crystals [19]. The search space is explored via a local search procedure. Simulated annealing escapes from local minima by allowing moves to worsening solutions with a probability that decreases in time.

Genetic algorithms are inspired by natural selection. In this metaheuristic, candidate solutions are represented as individuals of a populations that evolve in time under the effect of a number of operators including *crossover* and *mutation*, which mimic the effects of their natural counterparts [20].

Iterated local search is one of the simplest metaheuristics. It is based on the reiteration of a local search procedure: It explores the neighborhoods of different solutions obtained via successive perturbations [21].

Ant colony optimization is inspired by the foraging behavior of ants [2]. Solutions are sampled based on a *pheromone* model and are used to modify the model itself biasing the search toward high quality solutions [22].

Each metaheuristic can be seen as a modular structure coming with a set of components, each typically provided with a set of free parameters. The tuning problem is the problem of properly instantiating this algorithmic template by choosing the best among the set of possible components and by assigning specific values to all free parameters [5]. Only in recent years this problem has been the object of extensive studies [5,23,24,25,26], although it is generally recognized to be very important when dealing with metaheuristics. Some authors adopt a methodology based on factorial design, which is characteristic of a *descriptive* analysis. For example, [27] try to identify the relative contribution of five different components of a tabu-search. Furthermore, the authors consider different values of the parameters of the most effective components and select the best one. [28] and [29] use a similar approach. [30] describe a more general technique, which is nonetheless based on factorial analysis. Another approach to tuning that has been adopted for example by [26] and by [23] is based on the method that in the statistical literature is known as *response surface methodology*. [25] propose a method to determine relevant parameter settings. Some procedures for tackling the tuning problem have been proposed by [5].

3 Main Elements of the Analysis

The three main elements of the case study considered in the paper are presented in this section.

3.1 The Problem

The VRPSD is addressed by considering only one vehicle [3,11,12,13]. An *a priori* TSP-wise tour is constructed and is then split according to the specific realizations of the demand of the customers. The objective is finding the *a priori* tour with minimum expected cost. The computation of the expected cost of solutions is based on a dynamic programming recursion that moves backward from the last node of the sequence. At each node, the decision of restocking or proceeding is based on the expected cost-to-go in the two cases [13,3].

Two local search procedures are considered: Or-opt and 3-opt [13,3]. Five methods are used for computing the cost of a move in the local search: Or-opt(TSP-cost), Or-opt(VRPSD-cost), Or-opt(EXACT-cost), 3-opt(TSP-cost), 3-opt(EXACT-cost). For a detailed description of these techniques we refer the reader to [3].

A rather large set of instances is needed in order to reach some significant conclusion with our empirical analysis. The set of instances considered in [3] is too small for the aim of our research. To the best of our knowledge, these are the only benchmark instances available for the vehicle routing problem with stochastic demand. For our experiments, we use instances created with the instance generator described in [31]. We consider instances with either 50 or 60 nodes.

Following [3], we consider instances in which the demand of each customer is uniformly distributed. The average and the spread of these distributions are selected randomly extracted from a uniform distribution in the following ranges: $\{(20, 30), (20, 35)\}$ for the average, and $\{(5, 10), (5, 15)\}$ for the spread. The capacity of the vehicle is 80.

3.2 Metaheuristics

The implementation of the metaheuristics we consider is based on the code written for [3], which is available at http://iridia.ulb.ac.be/vrpsd.ppsn8. In the following, we give a short description of the main element characterizing each algorithm. More details can be found in [32]. The parameters of the algorithms are briefly explained. As a reference algorithm, following [3], we considered a random restart local search (RR). It uses the randomized furthest insertion heuristic plus local search. It restarts every time a local optimum is found, until the stopping criterion is met—in our case, the elapsing of a fixed computational time.

In **tabu search**, the tabu-list stores partial solutions. An *aspiration criterion* allows forbidden moves if the new solution is the new best one. The *tabu tenure*, that is, the length of the tabu list, is variable [3]: At each step it assumes a random value between $t(m-1)$ and $m-1$, where $0 \leq t \leq 1$ is a parameter of the algorithm. When 3-opt is used, m is equal to the number of customers. When Or-opt is used, m is equal to the number of customers minus the length of the string to move. During the exploration of the neighborhood, solutions that include forbidden components are evaluated with probability p_f and the others with probability p_a. The difference between the EXACT-cost, the VRPSD-cost, and the TSP-cost implementations concerns only the local search procedure.

Concerning **simulated annealing**, the probabilistic acceptance criterion consists in accepting a solution s' either if it has a lower cost than the current solution s or, independently of its cost, with probability $p(s'|T_k, s) = exp(Cost(s) - Cost(s')/T_k)$. The relevant parameters of the algorithm are related to the initial level of the *temperature* and to its evolution. The starting value T_0 is determined by considering one hundred solutions randomly chosen in the neighborhood of the first one, by computing the variation of the cost in this set, and by multiplying this result for the parameter f. At every iteration k, the *temperature* is decreased according to the formula $T_k = \alpha T_{k-1}$, where the parameter α, usually called *cooling rate*, is such that $0 < \alpha < 1$. If after $n \cdot q \cdot r$ iterations the quality of the best solution is not improved, the process known as *re-heating* [33] is applied: the temperature is increased by adding T_0 to the current temperature. Besides the local search procedure used, the difference between the EXACT-cost, the VRPSD-cost and the TSP-cost implementations consists in the way $Cost(s')$ and $Cost(s)$ are computed. In the TSP-cost, only the length of the *a priori* tour is considered.

In the implementation of **genetic algorithm**, edge recombination [34] consists in generating a tour starting from two solutions by using edges present in both of them, whenever possible. Mutation swaps adjacent customers with probability p_m. If mutation is *adaptive*, p_m is equal to the product of the parameter mr (*mutation-rate*) and a similarity factor. The latter depends on the number of times the n-th element of the first parent is equal to the n-th element of the second one. If the mutation is not *adaptive*, p_m is simply equal to mr. The difference between the EXACT-cost, the VRPSD-cost and the TSP-cost implementations concerns only the local search procedure adopted.

Iterated local search is characterized by a function that performs a perturbation on solutions. It returns a new solution obtained after a loop of n random moves (with n number of nodes of the graph) of a 2-exchange neighborhood. They consist in subtour inversions between two randomly chosen nodes. The loop is broken if a solution with quality comparable to the current one is found. We say that the quality of a solution is comparable to the quality of the current one if its objective function value is not greater than the objective function value of the current solution plus a certain value ϵ. The difference between the EXACT-cost, the VRPSD-cost and the TSP-cost implementations concerns only the local search procedure adopted.

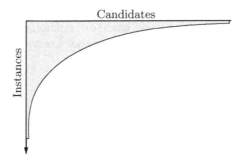

Fig. 1. Graphical representation of computation performed by the racing approach. As the evaluation proceeds, the racing algorithm focuses more and more on the most promising candidates, discarding a configuration, as soon as sufficient evidence is gathered that it is suboptimal [5].

In this implementation of **ant colony optimization**, the pheromone trail is initialized to $\tau_0 = 0.5$ on every arc. The first population of solutions is generated and refined via the local search. Then, a *global pheromone update* is performed r times. At each following iteration, p new solutions are constructed by p artificial ants on the basis of the information stored in the pheromone matrix. After each step, the *local pheromone update* is performed on the arc just included in the route. Finally, the local search is applied to the p solutions and the *global pheromone update* is executed.

Local pheromone update: the pheromone trail on the arc (i, j) is modified according to $\tau_{ij} = (1 - \psi)\tau_{ij} + \psi\tau_0$, with ψ parameter such that $0 < \psi < 1$.

Global pheromone update: the pheromone trail on each arc (i, j) is modified according to $\tau_{ij} = (1 - \rho)\tau + \rho\Delta\tau_{ij}^{bs}$ where $\Delta\tau_{ij}^{bs} = Q/Cost\_Solution\_bs$ if arc (i, j) belongs to *Solution_bs*, and $\Delta\tau_{ij}^{bs} = 0$ otherwise. ρ is a parameter such that $0 < \rho < 1$ and *Solution_bs* is the best solution found so far.

3.3 The Tuning Process

The parameters of all algorithms considered in the paper are tuned through the F-Race procedure [5,4]. F-Race is a racing algorithm for choosing a candidate configuration, that is, a combination of values of the parameters, out of predefined ranges. A racing algorithm consists in generating a sequence of nested sets of candidate configurations to be considered at each step (Figure 1). The set considered at a specific step h is obtained by possibly discarding from the set considered at step $h - 1$, some configurations that appear to be suboptimal on the basis of the available information. This cumulated knowledge is represented by the behavior of the algorithm for which the tuning is performed, when using different candidates configurations. For each instance (each representing one step of the race) the ranking of the results obtained using the different configurations is computed and a statistical test is performed for deciding whether to discard some candidates from the following experiments (in case they appear suboptimal) or not. F-Race is based on the Friedman two-way analysis of variance by ranks [35]. An important advantage offered by this statistical test is connected with the nonparametric nature of a test based on ranking, which does not require to formulate hypothesis on the distribution of the observations.

4   Experimental Analysis

With the computational experiments proposed in this section we wish to show that a remarkable difference exists between the results obtained by *out-of-the-box* and *custom* versions of metaheuristics.

As mentioned in the introduction, the two versions of the metaheuristics are obtained by applying different procedures for setting the values of the parameters. The values of the parameters represent only one of the elements that may be customized when implementing a metaheuristic. In this sense, here the difference between *out-of-the-box* and *custom* implementations may be underestimated. Then, our goal is reached by showing that the results achieved with *out-of-the-box* implementations cannot be generalized to *custom* ones even when the difference between the two simply consists in this element: Any other element that can be fine-tuned and customized would simply further reduce the possibility of generalizing results observed in one context to the other.

In the *custom* versions, the parameters are accurately fine-tuned with the F-Race automatic procedure. In the *out-of-the-box* versions, the values of the parameters are randomly drawn from the same set of candidate values that is considered by F-Race for *custom* versions. Equal probability has been associated to each configuration and, for each instance considered in the analysis, a random selection has been performed.

For each of the metaheuristics, besides the methods used for setting the parameters, the implementations considered in the *out-of-the-box* and *custom* versions are identical.

All experiments are run on a cluster of AMD Opteron™ 244, and 1000 instances are considered. A computation time of 30 seconds is used as a stopping criterion for all the algorithms.

In order to obtain the *custom* versions of the metaheuristics through F-Race, a number of different configurations ranging from 1200 to about 1600 were considered for each of them. Table 1 reports, for each metaheuristic, the parameters that have been considered for optimization, the range of values allowed, and the values that have been selected. A set of 500 instances of the vehicle routing problem with stochastic demand was available for the tuning. These instances have the same characteristics of the ones used for the experimental analysis, but the two sets of instances are disjoint [36]. While tuning a metaheuristic, the F-Race procedure was allowed to run the metaheuristic under consideration for a maximum number of times equal to 15 times the number of configurations considered for that metaheuristic. Also for the random restart local search, a *custom* version has been considered. It has been obtained by selecting, through the F-Race procedure, the best performing local search. In other words, the parameter that has been optimized in this case is the underlying local search.

Table 1. Range of values considered for the parameters of the metaheuristics. The values reported in bold are the ones selected by F-Race for the *custom* versions.

Tabu search – total number of candidates = 1460	
parameter	range
p_f	0.1, **0.2**, 0.25, 0.3, 0.35, 0.4
p_a	0.5, **0.6**, 0.7, 0.75, 0.8, 0.85, 0.9
t	**0.3**, 0.4, 0.5, 0.7, 0.8, 0.9, 1
local_search	Or-opt(TSP-cost), Or-opt(VRPSD-cost), Or-opt(EXACT-cost), 3-opt(TSP-cost), **3-opt(EXACT-cost)**

Simulated annealing – total number of candidates = 1200	
parameter	range
α	**0.3**, 0.5, 0.7, 0.9, 0.98
q	**1**, 5, 10
r	10, 20, 30, **40**
f	0.01, **0.03**, 0.05, 0.07
local_search	**Or-opt(TSP-cost)**, Or-opt(VRPSD-cost), Or-opt(EXACT-cost), 3-opt(TSP-cost), 3-opt(EXACT-cost)

Genetic algorithm – total number of candidates = 1360	
parameter	range
pop. size	10, 12, 14, 16, 18, **20**, 22, 24
mr	0.1, 0.15, 0.2, 0.25, 0.3, 0.35, 0.4, 0.45, 0.5, 0.55, 0.6, 0.65,**0.7**, 0.75, 0.8, 0.85, 0.9
adaptive	**Yes**, No
local_search	**Or-opt(TSP-cost)**, Or-opt(VRPSD-cost), Or-opt(EXACT-cost), 3-opt(TSP-cost), 3-opt(EXACT-cost)

Iterated local search – total number of candidates = 1520	
parameter	range
ϵ	$n/x, x \in \{0.005, 0.01, 0.05, 0.1, 0.5, 1.0, \mathbf{1.5}, 2.0$, all multiples of 0.5 up to 150.0 $\}$
local_search	**Or-opt(TSP-cost)**, Or-opt(VRPSD-cost), Or-opt(EXACT-cost), 3-opt(TSP-cost), 3-opt(EXACT-cost)

Ant colony optimization – total number of candidates = 1620	
parameter	range
p	**5**,10, 20
ρ	0.1, 0.5, **0.7**
r	100, **150**, 200
Q	$10^5, 10^6, 10^7, \mathbf{10^8}, 10^9$
local_search	Or-opt(TSP-cost), Or-opt(VRPSD-cost), Or-opt(EXACT-cost), **3-opt(TSP-cost)**, 3-opt(EXACT-cost)

Random restart – total number of candidates = 5	
parameter	range
local_search	Or-opt(TSP-cost), Or-opt(VRPSD-cost), Or-opt(EXACT-cost), 3-opt(TSP-cost), **3-opt(EXACT-cost)**

First of all, let us compare the results achieved by the metaheuristics in the two contexts in terms of cost of the best solution returned. The whole distribution of the difference of the results is reported in Figure 2 for each metaheuristic. The detail of the region around 0 is presented in Figure 2(b). The cost of the solutions found by each *custom* version minus the one of its *out-of-the-box* counterpart is considered for each instance. Even if the tails of the distributions are sometimes very long, it can be observed that almost 75% of the observations fall below the zero line for all metaheuristics: the difference is in favor of the *custom* version in the strong majority of the cases. Moreover, it can be noted that, as it can be expected, the various metaheuristics are sensitive in different measure to the value of their parameters. Therefore, they may benefit in different measure from an accurate fine-tuning. Observing these results, it is immediately clear that, as

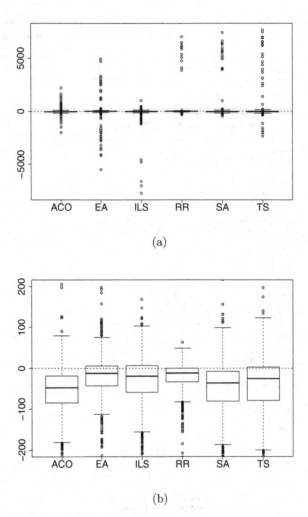

Fig. 2. Difference between the costs of the solutions obtained by the *custom* and the *out-of-the-box* versions of the metaheuristics under analysis. In Figure 2(a), the entire distribution is shown for each metaheuristic. Since the distributions are characterized by long tails, in Figure 2(b) the detail of the more interesting central area is given. For all metaheuristics, the median of the distribution is below the zero, which means that the results obtained by the *custom* versions are in general better than those obtained by their *out-of-the-box* counterpart.

expected, the performance achieved by algorithms depend strongly on the values chosen for the parameters, and then on the contexts considered.

Some further observations can be made considering the distribution of the ranking achieved by each algorithm. Figures 3(a) and 3(b) report the results achieved by the *custom* and *out-of-the-box* versions, respectively. On the left of each graph, the names of the algorithms are given. The order in which they

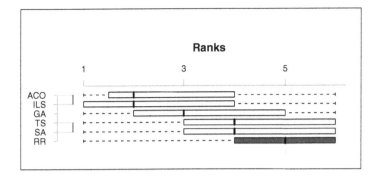

(a) *Custom* versions

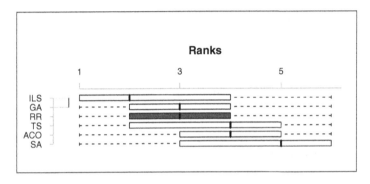

(b) *Out-of-the-box* versions

Fig. 3. Results over 1000 instances of the metaheuristics in the two variants considered

appear reflects the average ranking: The lower the average ranking, the better the general behavior, and the higher the metaheuristic appears in the list. On the right, the boxplots represent the distributions of the ranks over the 1000 instances. Between the names and the boxplots, vertical lines indicate if the difference in the behavior of the metaheuristics is significant according to the Friedman test: If two metaheuristics are not comprised by the same vertical line, their behavior is significantly different according to the statistical test considered, with a confidence of 95%.

As it can be observed, the ranking of algorithms varies in the two contexts: The two main differences concern RR and ACO. The former performs the worst in the *custom* context, while this is not the case in the *out-of-the-box* context. The case of a metaheuristic performing worse than the random restart local search is to be considered as a major failure for the metaheuristic itself. We consider this point as a remarkable difference between the two contexts: In the *out-of-the-box* context, three out of five metaheuristics perform significantly worse than the random restart local search; in the *custom* context, all metaheuristics achieve better results than the random restart local search.

As far as ACO is concerned, we can draw a conclusion that was suggested by the representation proposed in Figure 2 and 2(b). In fact, it can observe that the relative performance is visibly different in the two contexts. The *out-of-the-box* version behaves significantly worse than RR, and is among the worst in the set. On the contrary, the *custom* version achieves the best average ranking. This difference shows that this metaheuristic is more sensitive than the others to variations of the parameters, possibly due to the large number of parameters of the algorithm. This might be seen as a drawback of ACO. Anyway, we think that this fact should be read in a different way: If one is interested in an *out-of-the-box* metaheuristics, a high sensitivity to the parameters is definitely an issue; on the other hand, if one wishes to implement a *custom* metaheuristic, the sensitivity is an opportunity that can be exploited in order to finely adapt the algorithm to the class of instances to be tackled.

Another point that can be observed concerns the comparison of the ranking obtained by the metaheuristics in the two contexts considered, and the one proposed in [3] on similar instances. Even if, certainly, more experiments are necessary before drawing conclusion, the general trend reported in [3] appears very similar to the one obtained in the *out-of-the-box* context.

These results clearly support our claim according to which there is a strong difference between the performance of metaheuristics used *out-of-the-box* or in a *custom* way. Moreover, they make us to wonder whether the versions that can be found in the literature are necessarily to be considered *custom*, when applied to problem instances that differ from those considered in the original study.

5   Conclusions

In the paper, five of the most successful metaheuristics, namely tabu search, simulated annealing, genetic algorithm, iterated local search, and ant colony optimization, have been compared on the vehicle routing problem with stochastic demand. These five metaheuristics and this same optimization problem have been the focus of a research recently published by [3].

Each approach has been considered both in an *out-of-the-box* and in a *custom* version. The procedure used for choosing the values of the parameters is the element that differentiates a *custom* version of a metaheuristic from the corresponding *out-of-the-box* one: In the former, the parameters are fine-tuned through the F-Race algorithm. In the latter, they are drawn at random. Our goal is to highlight that results obtained in one context cannot be directly generalized to the other.

As it could be expected, the empirical results show that the *custom* version of each metaheuristic achieves better results than the corresponding *out-of-the-box* one. The difference is always statistically significant according to the Friedman test. Moreover, the relative performance of algorithms differs greatly in the two contexts. This can be ascribed to the fact that different metaheuristics might be more or less sensitive to variations of their parameters.

On the basis of this case study, we can conclude that there may be a strong difference in the results achievable by using the *out-of-the-box* or the *custom* version of metaheuristics. This difference may concern both the quality of the solutions returned by an approach, and the relative performance of algorithms. As a consequence, one should clearly describe the implementation criteria followed in the design of an algorithm, in order to allow the readers to focus on the more suitable implementations, given their specific goals.

The experimental analysis raises doubts on the possibility of *a priori* ascribing the results that are reported in the literature to one of the two contexts. The computations reported suggest that this is not the case. A further analysis needs to be devoted to this point.

Acknowledgments. This work was supported by the ANTS project, an *Action de Recherche Concertée* funded by the Scientific Research Directorate of the French Community of Belgium.

References

1. Glover, F.: Future paths for integer programming and links to artificial intelligence. Computers & Operations Research 13, 533–549 (1986)
2. Dorigo, M., Stützle, T.: Ant Colony Optimization. MIT Press, Cambridge, MA, USA (2004)
3. Bianchi, L., Birattari, M., Chiarandini, M., Manfrin, M., Mastrolilli, M., Paquete, L., Rossi-Doria, O., Schiavinotto, T.: Hybrid metaheuristics for the vehicle routing problem with stochastic demands. Journal of Mathematical Modelling and Algorithms 5(1), 91–110 (2006)
4. Birattari, M., Stützle, T., Paquete, L., Varrentrapp, K.: A racing algorithm for configuring metaheuristics. In: Langdon, W. (ed.) GECCO 2002: Proceedings of the Genetic and Evolutionary Computation Conference, pp. 11–18. Morgan Kaufmann, San Francisco (2002)
5. Birattari, M.: The problem of tuning metaheuristics as seen from a machine learning perspective. PhD thesis, Université Libre de Bruxelles, Brussels, Belgium (2005)
6. Bartz-Beielstein, T.: Experimental analysis of evolution strategies - overview and comprehensive introduction. Technical Report CI-157/03, Interner Bericht des Sonderforschungsbereichs 531 Computational Intelligence, Universität Dortmund, Dortmund, Germany (2003)
7. Tillman, F.: The multiple terminal delivery problem with probabilistic demands. Transportation Science 3, 192–204 (1969)
8. Stewart, W., Golden, B.: Stochastic vehicle routing: a comprehensive approach. European Journal of Operational Research 14, 371–385 (1983)
9. Dror, M., Trudeau, P.: Stochastic vehicle routing with modified saving algorithm. European Journal of Operational Research 23, 228–235 (1986)
10. Laporte, G., Louveau, F., Mercure, H.: Models and exact solutions for a class of stochastic location-routing problems. Technical Report G-87-14, Ecole des Hautes Etudes Commerciale, University of Montreal, Montreal, Canada (1987)
11. Bertsimas, D.: A vehicle routing problem with stochastic demand. Operations Research 40(3), 574–585 (1992)

12. Bertsimas, D., Simchi-Levi, D.: A new generation of vehicle routing research: robust algorithms, addressing uncertainty. Operations Research 44(3), 286–304 (1996)
13. Yang, W., Mathur, K., Ballou, R.: Stochastic vehicle routing problem with restocking. Transportation Science 34(1), 99–112 (2000)
14. Secomandi, N.: A rollout policy for the vehicle routing problem with stochastic demands. Operations Research 49, 796–802 (2001)
15. Secomandi, N.: Analysis of a rollout approach to sequencing problems with stochastic routing applications. Journal of Heuristics 9, 321–352 (2003)
16. Teodorović, D., Pavković, G.: A simulated annealing technique approach to the VRP in the case of stochastic demand. Transportation Planning and Technology 16, 261–273 (1992)
17. Gendreau, M., Laporte, G., Séguin, R.: A tabu search heuristic for the vehicle routing problem with stochastic demands and customers. Working paper, CRT, University of Montreal, Montreal, Canada (1994)
18. Glover, F., Laguna, M.: Tabu Search. Kluwer Academic Publishers, Norwell, MA, USA (1997)
19. Ingber, L.: Adaptive simulated annealing (ASA): lessons learned. Control and Cybernetics 26(1), 33–54 (1996)
20. Bäck, T., Fogel, D., Michalewicz, Z. (eds.): Handbook of Evolutionary Computation. IOP Publishing Ltd. Bristol, UK (1997)
21. Lourenço, H., Martin, O., Stützle, T.: Iterated local search. In: Glover, F., Kochenberger, G. (eds.) Handbook of Metaheuristics, pp. 321–353. Kluwer Academic Publishers, Norwell, MA, USA (2002)
22. Zlochin, M., Birattari, M., Meuleau, N., Dorigo, M.: Model-based search for combinatorial optimization: A critical survey. Annals of Operations Research 131(1–4), 373–395 (2004)
23. Adenso-Díaz, B., Laguna, M.: Fine-tuning of algorithms using fractional experimental designs and local search. Operations Research 54(1), 99–114 (2006)
24. Barr, R., Kelly, J., Resende, M., Stewart, W.: Designing and reporting computational experiments with heuristic methods. Journal of Heuristics 1(1), 9–32 (1995)
25. Bartz-Beielstein, T., Markon, S.: Tuning search algorithms for real-world applications: A regression tree based approach. In: Greenwood, G. (ed.) Proc. 2004 Congress on Evolutionary Computation (CEC'04), Piscataway, NJ, USA, pp. 1111–1118. IEEE Computer Society Press, Los Alamitos (2004)
26. Coy, S., Golden, B., Runger, G., Wasil, E.: Using experimental design to find effective parameter settings for heuristics. Journal of Heuristics 7(1), 77–97 (2001)
27. Xu, J., Kelly, J.: A network flow-based tabu search heuristic for the vehicle routing problem. Transportation Science 30, 379–393 (1996)
28. Parson, R., Johnson, M.: A case study in experimental design applied to genetic algorithms with applications to dna sequence assembly. American Journal of Mathematical and Management Sciences 17, 369–396 (1997)
29. Breedam, A.V.: An analysis od the effect of local improvement operators in genetic algorithms and simulated annealing for the vehicle routing problem. Technical Report TR 96/14, Faculty of Applied Economics, University of Antwerp, Antwerp, Belgium (1996)
30. Xu, J., Chiu, S., Glover, F.: Fine-tuning a tabu search algorithm with statistical tests. International Transactions on Operational Research 5(3), 233–244 (1998)
31. Pellegrini, P., Birattari, M.: Instances generator for the vehicle routing problem with stochastic demand. Technical Report TR/IRIDIA/2005-10, IRIDIA, Université Libre de Bruxelles, Brussels, Belgium (2005)

32. Pellegrini, P., Birattari, M.: The relevance of tuning the parameters of metaheuristics. A case study: The vehicle routing problem with stochastic demand. Technical Report TR/IRIDIA/2006-008, IRIDIA, Université Libre de Bruxelles, Brussels, Belgium (submitted for journal publication, 2006)
33. Aarts, E., Korst, J., van Laarhoven, P.: Simulated annealing. In: Aarts, E., Lenstra, J. (eds.) Local Search in Combinatorial Optimization, pp. 91–120. John Wiley & Sons, Inc. New York, USA (1997)
34. Whitley, D., Starkweather, T., Shaner, D.: The traveling salesman problem and sequence scheduling: quality solutions using genetic edge recombination. In: Davis, L. (ed.) Handbook of Genetic Algorithms, pp. 350–372. Van Nostrand Reinhold, New York, USA (1991)
35. Friedman, J.: Multivariate adaptive regression splines. The Annals of Statistics 19, 1–141 (1991)
36. Birattari, M., Zlochin, M., Dorigo, M.: Towards a theory of practice in metaheuristics design: A machine learning perspective. Theoretical Informatics and Applications, Accepted for publication (2006)

Tuning the Performance of the MMAS Heuristic

Enda Ridge and Daniel Kudenko

Department of Computer Science, The University of York, England
Enda.Ridge@googlemail.com, Kudenko@cs.york.ac.uk

Abstract. This paper presents an in-depth Design of Experiments (DOE) methodology for the performance analysis of a stochastic heuristic. The heuristic under investigation is Max-Min Ant System (MMAS) for the Travelling Salesperson Problem (TSP). Specifically, the Response Surface Methodology is used to model and tune MMAS performance with regard to 10 tuning parameters, 2 problem characteristics and 2 performance metrics—solution quality and solution time. The accuracy of these predictions is methodically verified in a separate series of confirmation experiments. The two conflicting responses are simultaneously optimised using desirability functions. Recommendations on optimal parameter settings are made. The optimal parameters are methodically verified. The large number of degrees-of-freedom in the MMAS design are overcome with a Minimum Run Resolution V design. Publicly available algorithm and problem generator implementations are used throughout. The paper should therefore serve as an illustrative case study of the principled engineering of a stochastic heuristic.

1 Introduction and Motivation

The MMAS heuristic is a member of the Ant Colony Optimisation (ACO) meta-heuristic family [1]. As such, it is a general purpose *approximate* algorithm for solving a variety of related problems. This generality comes at a price. Typical of most heuristics, ACO algorithms are stochastic and have many possible tuning parameters that influence their performance. Performance is further influenced by one or more problem instance characteristics. This presents several problems for the heuristic engineer. For the heuristic to be of any practical use, the engineer must be able to:

- Determine the tuning parameters and problem characteristics affecting performance.
- Relate these parameters and characteristics to performance so that the best parameter settings can be chosen for given problem instance characteristics.
- Allow for the fact that stochastic heuristics will vary in performance between identical repetitions of the same run.
- Perform a simultaneous analysis of at least two, often conflicting, responses.

This is the *parameter tuning problem*. Despite the parameter tuning problem being central to our understanding of heuristic performance, its treatment in

T. Stützle, M. Birattari, and H.H. Hoos (Eds.): SLS 2007, LNCS 4638, pp. 46–60, 2007.

the literature is lacking. Existing work has indeed provided useful insights into the likely importance and recommended settings of some tuning parameters. However, research could be improved by addressing the following concerns.

- Research rarely reports whether parameters were *methodically* chosen and how such a choice was made. This leaves the reader with no knowledge of how to repeat the procedure themselves, if indeed any recognised methodology was used.
- Research often uses a single performance measure. The stochastic and approximate nature of heuristics necessitates that we simultaneously consider at least two conflicting performance measures.

Researchers that do attempt to methodically tune parameters, often do so with a One-Factor-At-A-Time (OFAT) approach. This approach examines the influence of a single tuning parameter value while holding all the other parameters constant. This paper demonstrates that a well-established methodical approach, called Design of Experiments (DOE), can successfully be brought to bear on the parameter tuning problem. The advantages of Design of Experiments over the OFAT approach are well recognised in other fields [2]:

- Designed experiments require fewer resources, in terms of experiments, time and material, for the amount of information obtained.
- Designed experiments can estimate interactions between factors[1]. This is not the case with OFAT experiments.
- There is experimental information in a larger region of the experiment design space. This makes process optimisation more efficient because the whole factor space can be searched.

The paper begins with a background on the MMAS algorithm and TSP problem domain, introducing and explaining the tuning parameters that will be examined. The DOE approach is also discussed. Section 3 presents the methodology. Sections 4, 5 and 6 describe experiment results. The paper then summarises some related work and presents our conclusions.

2   Background

2.1   The MMAS Heuristic for the TSP

Max-Min Ant System (MMAS) [3] is a member of the Ant Colony Optimization (ACO) metaheuristic family. In this research, MMAS is applied to the Travelling Salesperson Problem (TSP). MMAS has been shown to be one of most effective ACO algorithms for the TSP [3].

---

[1] A *factor* is an independent variable manipulated in an experiment. Here, a factor describes both the tuning parameters and problem characteristics that we manipulate in the experiments.

The TSP is the problem of visiting a set of cities in the shortest distance possible such that each city is visited only once. It is a *discrete combinatorial optimisation* problem. The TSP is often represented by a graph data structure where nodes in the graph represent cities and edges represent the cost of moving between cities.

Broadly, ACO is based on a metaphor of how natural ants forage for food. Artificial ants create tours of the TSP graph, depositing 'pheromone' along graph edges composing those tours. Ant movement decisions are probabilistic and are based on the pheromone levels on graph edges and a heuristic incorporating the cost of moving along an edge. Repeated iteration of the artificial ants' tour building and pheromone deposition and decay allow ACO algorithms to converge on solutions to the optimisation problem.

We summarise MMAS in 7 stages so that the reader can identify the tuning parameters and their intended affect on performance. More detailed descriptions are in the literature [1]. The MMAS implementation is a Java port of publicly available C code[2]. This port was informally verified to produce the same output as the original code given several sets of inputs and instances.

STAGE 1 Initialise pheromone: A starting pheromone value is assigned to all graph edges. In this case, all graph edges are set to

$$\frac{1}{\rho \cdot T_{nn}}$$

where ρ is a parameter discussed later and T_{nn} is the length of a nearest neighbour tour of the problem.

STAGE 2 Ant Placement: this involves ants being given a starting node on the graph. We explore two possibilities.

1. **Scatter:** all ants are scattered across randomly chosen nodes on the graph.
2. **Same node:** all ants begin at a single randomly chosen node.

STAGE 3 Tour construction: The ant movement decisions are either *explorative* or *exploitative*. A number is uniformly randomly chosen between 0.0 and 1.0. If this number is less than an exploration threshold q, the ant simply chooses the city with the next best combination of pheromone and edge cost. If the number is greater than or equal to the exploration threshold, then the ant makes a probabilistic decision on the next city to visit. The probability a city j will be chosen by an ant at city i is

$$p_{ij} = \frac{[\tau_{ij}]^\alpha [\eta_{ij}]^\beta}{\sum_{l \in F_i} [\tau_{il}]^\alpha [\eta_{il}]^\beta}, j \in F_i$$

where η_{ij} is the inverse of the cost of edge ij, F_i is the set of all unvisited (feasible) cites to visit after city i and α and β are parameters. The higher α

---

[2] http://iridia.ulb.ac.be/~mdorigo/ACO/aco-code/public-software.html

and β are, the greater is the influence of their associated terms on the choice of next city. The maximum and minimum limits on trail pheromone levels are then updated according to the following:

$$\tau_{max} = \frac{1}{\rho \cdot T_{best\_so\_far}} \text{ and } \tau_{min} = \frac{\tau_{max}}{2n}$$

where n is the problem size.

STAGE 4 Local search: it is common practice to improve the ants' tours with a local search procedure. Given the potentially large number of local search procedures that could be plugged into MMAS, we chose to experiment without any local search. However, the methodology and analysis presented here can incorporate investigating local search.

STAGE 5 Evaporate Pheromone: pheromone on all edges is evaporated according to $\tau_{ij} = (1 - \rho)\tau_{ij}$ where ρ is a tuning parameter.

STAGE 6 Trail update: Pheromone is then deposited along the trail of a single ant according to

$$\tau_{ij} = \tau_{ij} + {}^1\!/\text{length of tour}$$

The single ant used for updates is a choice between the best ant from the current iteration (iteration_best_ant) or the best ant in all iterations so far (best_so_far). We have encountered this choice incorporated into the MMAS logic in two different ways.

1. **Fixed frequency.** iteration_best_ant is used with an occasional use of best_so_far according to a fixed frequency.

2. **Schedule.** iteration_best_ant is used with an increasingly frequent use of best_so_far according to a schedule.

This research uses fixed frequency. Firstly, it is a simpler form of schedule which should be understood before experimenting with the more complicated schedule. Furthermore, as with the local search decision, there are many possible schedules that could be applied. Experimenting with all of these was beyond the scope of this work.

STAGE 7 Pheromone reinitialisation: Pheromone reinitialisation is triggered when the algorithm is stagnating. The algorithm judges that it is stagnating when *both* of two conditions are met[3]. Firstly, the variation in pheromone levels (called *branching factor*) on edges drops below a threshold. Secondly, a number of iterations since the last improvement is exceeded. In the original source code, this check was made with a given frequency of iterations. We were

---

[3] The literature cites that one condition *OR* another has been met .[1, p. 76]. This research remains compatible with the original source code and uses an *AND* condition.

concerned that this would mask the iteration threshold condition so our implementation checks for reinitialisation after every iteration. If a reinitialisation is due, trails are reset to $\tau_{\max}$.

Further comments. Several further parameters are implicit in the MMAS implementation. The calculation of branching factor is the number of edges whose pheromone levels exceed the cutoff level given by min + λ(max - min) where max and min are the maximum and minimum pheromone values on all edges. This obviously contains a parameter λ which, in this research and the original source code, was fixed at 0.05.

Ant decisions in tour construction are limited to a *candidate list* of edges due to the computational expense of evaluating all edges. A candidate list is an ordered list of the nearest cities to a given city. The computationally expensive evaluation of branching factor is also limited in this way.

2.2   Experiment Designs for Response Surface Models

The overall goal of this research is to investigate how changes in algorithm tuning parameters and problem characteristics affect algorithm performance. The tuning parameters and characteristics we vary are called *factors* and the performance measures we gather are called *responses*. When investigating the effect of factors on some response(s), the general approach is to vary those factors over some number of *levels*. The experimental region covered by our chosen ranges of factors is called the *design space*. We measure the responses and interpolate these measurements into a *response surface* over the design space. This is the *Response Surface Methodology* (RSM) [4]. One of the most popular families of designs for building response surfaces are known as the *Central Composite Designs* (CCD). A CCD contains an embedded factorial design[4] augmented with design points at the centre of the design space (*centre points*) and so-called *axial points* located at some distance α from the design centre. The two most popular CCD designs are the Circumscribed Central Composite and Face-Centred Central Composite.

The *Circumscribed Central Composite* (CCC) sets α at such a value that the axial points create a circle that circumscribes the square defined by the embedded factorial. The *Face-Centered Composite* (FCC) sets α such that the axial points lie on the faces of the square defined by the embedded factorial. This research uses a Face-Centred Composite design because of the practical limits of parameters such as the exploration threshold of Section 2.1 where, for example, $0.0 \leq q \leq 1.0$. Detailed descriptions of CCC and FCD are in the literature [4].

The use of a full factorial in the embedded part of a CCD is expensive. A crossing of say 12 factors (as needed in this research), each at the minimum of 2 levels, would require $2^{12} = 4096$ design points for the embedded factorial alone. Recently, a state-of-the-art design called a *Minimum Run Resolution V* design has been introduced [5]. This provides a vast saving in experiment runs while still allowing all main and second-order interactions to be estimated.

---

[4] A full factorial design crosses all levels of all factors with one another.

3 Methodology

3.1 Stagnation Stopping Criterion

This research takes a practical view that once an algorithm is no longer producing improved results regularly, it is preferable to halt its execution and better employ the computational resources. This leads to using *stagnation* as a stopping criterion where stagnation is a number of iterations in which no solution improvement is observed. Stagnation offers the reproducibility of a combinatorial count while incorporating problem difficulty into the algorithm's halting. It also avoids the use of CPU time as a stopping criterion. This has been criticised as 'not acceptable for a scientific paper' [6] on the grounds of reproducibility. This research uses 250 iterations. Of course, results are dependent on the criterion chosen. This is the nature of experiments with heuristics.

3.2 Responses

Performance measures must reflect the conflicting heuristic goals of high solution quality and low time to solution. Johnson advocates reporting running times.

> a key reason for using approximation algorithms is to trade quality of solution for reduced running time, and readers may legitimately want to know how that tradeoff works out for the algorithms in question. [6]

This research specifically addressed this trade-off. The response that reflects time-to-solution is the CPU time from the beginning of an algorithm run to the time that the algorithm has stagnated and halted, *not* to the time the best solution was found. We were careful not to include time to calculate and write output that is not essential to the algorithm's functioning.

It is an open question as to which solution quality response is the more appropriate. This research uses *relative error*, defined as the difference between the algorithm solution and the optimal solution expressed as a percentage of the optimal solution. Alternatives such as Adjusted Differential Approximation [7] also exist. The optimal was calculated using the Concorde solver [8].

3.3 Problem Instances

Since this research includes problem size and standard deviation of problem edge cost as factors, we need a method to produce instances with controllable levels of these characteristics. Publicly available benchmark libraries do not have the breadth to provide such instances. For reproducibility, we used our own Java implementation of the 8th DIMACS Implementation Challenge [9] problem generator. This was informally verified to produce the same instances as the original C code. Our generator was then modified to draw its edge costs from a log-normal distribution where the standard deviation of the resulting edge costs could be controlled while their mean was fixed at 100. This was inspired by previous investigations on problem difficulty for an exact TSP algorithm [10] and verified for the MMAS heuristic [11].

3.4   Experiment Design

The reported research used a Minimum Run Resolution V Face-Centred Composite design with 8 replicates of the factorial and axial points and 6 centre points. This requires a total of 1452 runs. A 5% statistical significance level was used throughout.

Table 1 lists the 12 experiment factors and their high and low factorial levels. Factor levels were chosen using two criteria. If the factor has bounds on its values (for example, $0 \leq q_0 \leq 1.0$) then these bounds were chosen as the high and low factor levels. Alternatively, if the factor has no bounds (for example, $0 < \alpha \leq \infty$), we chose high and low values that incorporate values typically seen in the literature [3,1].

Table 1. Experiment factors and high (+) and low (-) factor levels. N denotes a numeric factor and C a categoric factor.

	Factor	Meaning	Type	(-)	(+)
A	Alpha	Exponentiates the pheromone term.	N	1.00	13.00
B	Beta	Exponentiates the distance heuristic term.	N	1.00	13.00
C	Ants	Number of ants expressed as % of problem size.	N	1.00	110.00
D	NNAnts	Length of candidate list as % of problem size.	N	1.00	20.00
E	q0	Exploration, exploitation threshold.	N	0.01	0.99
F	Rho	Phermone evaporation parameter.	N	0.01	0.99
G	Branch	The branching factor threshold below which a trail reinitialisation occurs.	N	0.5	2.0
H	Iters	The iteration threshold after which a trail reinitialisation occurs.	N	2	8
J	StDev	Standard deviation of edges in the TSP	N	10.00	70.00
K	Size	Number of cities in the TSP instance.	N	300	500
L	Best_so_far Freq	The iteration frequency with which best_so_far ant is used for pheromone updates.	N	2	40
M	Ant Place- ment	Whether ants are scattered randomly across the graph or at the single randomly chosen city.	C	random	same

The RSM combinations of factors J and K required 9 problem instances (including the centre points). Since experiments were conducted on 5 similar (but not identical) machines, it was necessary to randomize the run order to deal with unknown and uncontrollable nuisance factors [12]. Such factors might include hardware differences and operating system differences that could impact on the CPU time response.

4   Model Fitting

41 outliers (3% of the data) were removed[5] to make the data amenable to statistical analysis. Outliers were chosen using the usual diagnostics[6]. All responses were transformed with a $\log_{10}$ transformation so that they then passed these tests satisfactorily. However, the normal plots of the solution quality responses exhibited more deviation from the line than previously encountered with another Ant Colony algorithm [13]. This may be due to the restart nature of MMAS. We pursued our analysis on the assumption that the ANOVA tests are robust to this small deviation and verified our model as detailed in Section 5.

A *fit analysis* was conducted for each response to determine the highest order statistically significant and unaliased model that could be fit to the responses. A fit analysis begins by fitting the lowest order model, a linear model, to the response. The next higher order model is then fitted to the same response. If no additional terms from the higher order model are statistically significant, it is not necessary to use the higher order model. This procedure continues until the highest required order is found. The Minimum Run Resolution V design is aliased for cubic models. This leaves a linear model, a 2 factor interaction model and a quadratic model to be considered.

Based on these results, quadratic models were thus built for each response and for each combination of the categoric factor of ant placement (Table 1). Space restrictions prevent reproducing these models here. Before drawing conclusions from the resulting response surface models for each response, we must confirm the accuracy of the models.

5   Model Verification

A common approach in DOE [14] is to randomly sample points within the design space, run the actual process (in this case, the algorithm) at those points, and compare the model's predictions to the measurements from the randomly generated algorithm runs. We use two criteria upon which our satisfaction with the model (and thus confidence in its predictions) can be judged [13].

- **Conservative:** prefer models that provide consistently higher predictions of relative error and higher solution time than those actually observed.
- **Matching Trend:** prefer models that match the trends in algorithm performance. The prediction of the parameter combinations that give the best and

---

[5] We acknowledge that some authors question the deletion of outliers since outliers do represent real data and may reveal something about the process being studied.

[6] Normal plot of internally studentised residuals; Plot of internally studentised residuals against predicted value; Plot of internally studentised residuals against run number; Plot of predicted versus actual values; Plot of externally studentised residuals against run order; Plots of Leverage, DFFITS, DFBETAS and Cook's Distance. Please see NIST/SEMATECH e-Handbook of Statistical Methods (http://www.itl.nist.gov/div898/handbook/) for the meanings of these tests and their interpretation.

worst performance should match the actual parameter combinations that give the observed best and worst performance.

We randomly generated 50 new experiment runs from combinations of factors within the design space. For the randomly chosen combinations of problem characteristics, completely new instances were generated. Each experiment run was replicated 5 times. The measured responses were then compared to the models' 95% *prediction intervals*. A prediction interval [14, p. 394] is simply a range within which 95% of runs will fall. Figure 1 illustrates the results for the Time and Relative Error responses on new instances.

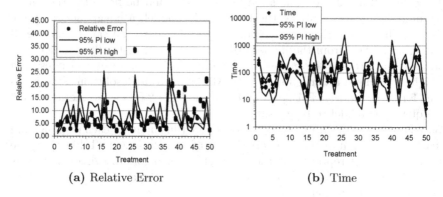

(a) Relative Error   (b) Time

Fig. 1. Verification of RSM predictions

The model is a better predictor of Time than Relative Error because of the small violations of the normality assumption in the Relative Error RSM. The Relative Error RSM does satisfy our matching trend criterion.

6   Response Optimisation

We express the multiple responses in terms of a single *desirability function* [15][7]. For each response $Y_i(x)$, its desirability function $d_i(Y_i)$ maps values between 0 to 1 to the possible values of $Y_i(x)$. $d_i(Y_i) = 0$ is completely undesirable and $d_i(Y_i) = 1$ is an ideal response. The *overall desirability* for all k responses is the geometric mean of the individual desirabilities:

$$D = \left(\prod_1^k d_i \right)^{\frac{1}{k}}$$

The contributions of each response to the overall desirability can be weighted to reflect user preferences. We use an equal weighting of quality and solution

---

[7] http://www.itl.nist.gov/div898/handbook/pri/section5/pri5322.htm

time. If the goal is to minimize a response (as in this research), the desirability functions take the following form where L_i, U_i and T_i are the lower bound, upper bound and target values of the desired response.

$$d_i\left(\hat{Y}_i\right) = \begin{cases} 1.0 & \text{if } \hat{Y}_i\left(x\right) < T \\ \left(\frac{\hat{Y}_i(x)-U_i}{T_i-U_i}\right) & \text{if } T_i \leq \hat{Y}_i\left(x\right) \leq U_i \\ 0 & \text{if } \hat{Y}_i\left(x\right) > U_i \end{cases}$$

Note that the fitted response value $\hat{Y}_i$ is used in place of Y_i. The Nelder-Mead downhill simplex [16, p. 326] is used to maximise the desirability of all responses. There are two important points to note about this optimisation:

- The two problem characteristics are factors in the model. If these were included in the optimisation, the optimisation would select a small problem size and low standard deviation. We therefore perform the optimizations with the two characteristics fixed at three-level factorial combinations.
- Recall that we are forcing Alpha and Beta to only take on integer values because of the expensive cost of non-integer exponentiation.

The results of these 9 optimizations are presented in Table 2 and a ranking of the relative size of contribution of each factor to the RSM models is presented in Table 3. *It is important to consider both optimisations and rankings since a factor that has a relative small effect on both responses can take on any value in the optimisation.* The following points should be noted.

- **Alpha and Beta:** these were generally kept low, except for large instances with high cost standard deviation.
- **Ants and Candidate List:** a small number of ants with a short candidate list was always preferred.
- **The exploration/exploitation threshold:** this is almost always at its maximum. This means exploration is rarely used and instead ant movement decisions are based on the best heuristic-product value.
- **Pheromone evaporation:** a higher value of ρ was preferred with increasing size and standard deviation.
- **Reinitialisation:** a high threshold iterations was preferred on smaller instances. A low branching factor threshold was always preferred. This is different from the value 1.0 that is usually fixed in the literature. Iteration threshold was one of the smallest contributors to both responses, suggesting that branching factor threshold is more important for determining trail reinitialisations.
- **Frequency of Best_so_far:** Smaller instances with lower standard deviation preferred a high frequency of use of best_so_far ant. Larger instances with higher standard deviations preferred to use best_so_far less frequently. However, this factor was one of the lower-ranking contributors to both responses (Table 3). This contradicts a result in the literature [1, p. 75].
- **Ant Placement:** this was one of the smallest contributors to both responses, suggesting that placement makes little difference to performance.

Table 2. Desirability optimisation results for Relative Error and Time on 9 combinations of problem size and problem standard deviation

size	StDev	alpha	beta	ants	NNAnts	q0	rho	Branch	Iters	Best_so_farFreq	Placement	Time	Rel Error	Desirability
300	10	1	1	1	1	0.99	0.01	0.5	80	2	random	0.5	0.2	1.0
300	40	6	1	1	1.22	0.99	0.84	0.6	78	23	random	0.7	0.7	0.9
300	70	9	1	1	1	0.97	0.60	1	80	36	same	0.7	1.5	0.8
400	10	1	1	1	1	0.99	0.01	0.5	80	2	same	1.3	0.5	0.9
400	40	1	1	1.1	1.44	0.99	0.92	1.08	70	28	random	2.9	1.3	0.8
400	70	12	11	1	2.58	0.01	0.72	0.51	80	37	same	1.8	2.3	0.7
500	10	4	3	1	1	0.99	0.45	0.5	22	23	random	3.2	0.7	0.8
500	40	11	1	1	1.09	0.99	0.7	0.5	40	39	same	3.6	2.4	0.7
500	70	13	13	9.1	1.13	0.01	0.79	0.5	43	40	same	4.1	2.1	0.7

Table 3. Ranks of contribution size of 12 factors for Relative Error and Time responses. The first two columns represent ranks within the main effects alone. The third and fourth columns represent ranks within all model effects.

	Time	Rel Error	Time	Rel Error
A-alpha	5	12	9	63
B-beta	12	3	64	3
C-ants	1	6	1	13
D-Candidate list	3	4	3	5
E-q0	4	2	4	2
F-rho	8	9	23	51
G-Branch Threshold	6	5	10	8
H-Iters Threshold	10	10	62	60
J-problemStDev	7	1	17	1
K-problemSize	2	7	2	27
L-Best_so_far Freq	9	8	50	29
M-ant Placement	11	11	63	62

7   Optimisation Verification

As with the predictions created with the RSMs, it is important to verify that the optimal parameter settings are indeed optimal and by how much they improve over other parameter settings. We randomly generate a set of TSP instances for each of the 9 factorial instance characteristic combinations of Section 6. We then run the MMAS algorithm with the optimal parameter settings on each of the 9 sets of instances. For each instance, we also run MMAS with 5 combinations of randomly chosen parameter settings. All runs were replicated 3 times. We should expect that for any instance, the optimal parameter settings produce solutions with lower solution time and lower relative error than the 5 randomly generated parameter settings. Figure 2 illustrates plots of the relative error and time respectively for 5 instances.

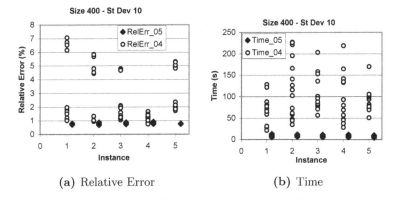

(a) Relative Error   (b) Time

Fig. 2. Comparison of optimal (shaded) and non-optimal (unshaded) parameter results

While some non-optimal results are close to the optimal results for relative error, none are lower than both relative error and time. A difficulty emerged with all instances that had a high standard deviation of cost matrix. Figure 3 illustrates how the relative error using non-optimal parameters was occasionally less than that using optimal parameters. However, the time results with optimal parameters remained less than the non-optimal parameters.

8   Related Work

Factorial designs have been combined with a local search procedure to systematically find the best parameter values for a heuristic [17]. Unfortunately CALIBRA can only tune five algorithm parameters. Its restrictive linear assumption precludes examining interactions between parameters. ACO algorithms require more than 5 parameters and our fitting analysis shows that interactions in higher order models are indeed important.

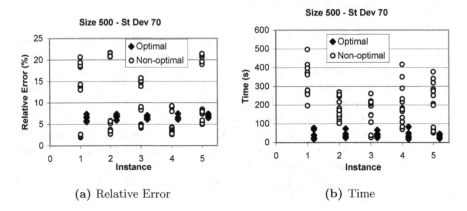

(a) Relative Error   (b) Time

Fig. 3. Comparison of optimal (shaded) and non-optimal (unshaded) parameter results

Coy *et al* [18] systematically find good settings for 6 tuning parameters on a set of Vehicle Routing Problems (VRP). Presented with a set of problems to solve, their procedure finds high quality parameter settings for a small number of problems in the problem set (the *analysis set*) and then combines these settings to achieve a good set of parameters for the complete problem set. A *fractional factorial design* is used to produce a response surface. The optimal parameter settings of the RSMs from the analysis set are averaged to obtain the final parameter settings for all problems in the set. Their method will perform poorly if the representative test problems are not chosen correctly or if the problem class is so broad that it requires very different parameter settings.

Park and Kim [19] used a non-linear response surface method to find parameter settings for a simulated annealing algorithm. Parsons and Johnson [20] used a central composite design with embedded fractional factorial to build a response surface and improve the performance of a genetic algorithm on a test data set. Only two parameters were modeled.

Birattari [21] uses algorithms derived from a machine learning technique known as *racing* to incrementally tune the parameters of several metaheuristics including Max-Min Ant System for the TSP [3].The author does not pursue the idea of a bi-objective optimisation of both time and quality. We have achieved this with a stagnation stopping criterion and the use of desirability functions.

A recent attempt to tune an ACO algorithm addressed only 3 parameters [22]. The authors then partitioned the three parameter ranges into 14, 9 and 11 values respectively. No reasoning was given for this granularity of partitioning or why the number of partitions varied between parameters. Each 'treatment' was run 10 times with a 1000 iteration or optimum found stopping criterion on a single 30 city instance. A single 30 city problem prevents an examination of problem characteristics and the size of 30 is so small as to be trivial for any algorithm. This resulted in 13,860 experiments. The approach was inefficient, requiring almost 10 times as many runs as this paper to tune one quarter as many parameters.

9   Contributions

The significance of the DOE techniques presented is as follows. We have been able to efficiently model and verify the performance of a stochastic heuristic in terms of two performance measures with a *Minimum Run Resolution V Face-Centred Composite Design*. There are several important contributions for the field of Ant Colony Optimization and the broader field of heuristics.

- **Efficient Experiment Designs.** Minimum Run Resolution V designs offer huge savings in experimental runs over full factorials and over fractional factorials. This saving overcomes the challenge of designing heuristics with a large number of degrees of freedom.
- **Simultaneous Analysis of Conflicting Responses.** Solution quality and solution time are critical to all heuristic performance analyses. *Desirability functions* permit simultaneously analysing and optimising these conflicting responses while assigning relative weights to each.
- **Optimal Parameter Recommendations.** A Nelder-Mead numerical optimisation of desirability allowed us to find parameter settings that optimise MMAS performance for different combinations of problem characteristics. The accuracy of these optimisations was demonstrated with an independent set of experiments. Several results were particularly noteworthy.

  - Higher values of alpha and beta are appropriate on some occasions.
  - A maximum exploration/exploitation threshold is almost always recommended, effectively ruling out the use of exploration in the algorithm. This is an unexpected result that merits further investigation.
  - Smaller instances with lower standard deviation preferred a high frequency of use of best_so_far ant. Larger instances with higher standard deviations preferred to use best_so_far less frequently. This contradicts a related result in the literature [1, p. 75].
  - A low threshold branching factor for trail reinitialisation was recommended. This was different from the fixed value of 1.0 used in the literature. We conclude that branching factor threshold should be considered as a tuning parameter.

A more detailed description of the DOE methodology and its adaptation to performance analysis of heuristics is available in the literature [23].

References

1. Dorigo, M., Stützle, T.: Ant Colony Optimization. MIT Press, Cambridge, MA (2004)
2. Czitrom, V.: One-Factor-at-a-Time versus Designed Experiments. The American Statistician 53(2), 126–131 (1999)
3. Stützle, T., Hoos, H.H.: Max-Min Ant System. Future Generation Computer Systems 16(8), 889–914 (2000)

4. Myers, R.H., Montgomery, D.C.: Response Surface Methodology. Process and Product Optimization Using Designed Experiments. John Wiley and Sons Inc. Chichester (1995)
5. Oehlert, G., Whitcomb, P.: Small, Efficient, Equireplicated Resolution V Fractions of 2K designs and their Application to Central Composite Designs. In: Proceedings of 46th Fall Technical Conference. American Statistical Association (2002)
6. Johnson, D.S.: A Theoretician's Guide to the Experimental Analysis of Algorithms. In: Proceedings of the Fifth and Sixth DIMACS Implementation Challenges (2002)
7. Zlochin, M., Dorigo, M.: Model based search for combinatorial optimization: a comparative study. In: Guervós, J.J.M., Adamidis, P.A., Beyer, H.-G., Fernández-Villacañas, J.-L., Schwefel, H.-P. (eds.) Parallel Problem Solving from Nature - PPSN VII. LNCS, vol. 2439, Springer, Heidelberg (2002)
8. Applegate, D., Bixby, R., Chvatal, V., Cook, W.: Implementing the Dantzig-Fulkerson-Johnson algorithm for large traveling salesman problems. Mathematical Programming Series B 97(1-2), 91–153 (2003)
9. Johnson, D.S., McGeoch, L.A.: Experimental analysis of heuristics for the STSP. In: The Traveling Salesman Problem and Its Variations, Kluwer Academic Publishers, Dordrecht (2002)
10. Cheeseman, P., Kanefsky, B., Taylor, W.M.: Where the Really Hard Problems Are. In: Proceedings of the Twelfth International Joint Conference on Artificial Intelligence, vol. 1, pp. 331–337. Morgan Kaufman, USA (1991)
11. Ridge, E., Kudenko, D.: An Analysis of Problem Difficulty for a Class of Optimisation Heuristics. In: Proceedings of EvoCOP '07. LNCS, vol. 4446, pp. 198–209. Springer, Heidelberg (2007)
12. Ostle, B.: Statistics in Research, 2nd edn. Iowa State University Press (1963)
13. Ridge, E., Kudenko, D.: Analyzing Heuristic Performance with Response Surface Models: Prediction, Optimization and Robustness. In: Proceedings of the Genetic and Evolutionary Computation Conference, ACM Press, New York (2007)
14. Montgomery, D.C.: Design and Analysis of Experiments, 6th edn. Wiley (2005)
15. Derringer, G., Suich, R.: Simultaneous Optimization of Several Response Variables. Journal of Quality Technology 12(4), 214–219 (1980)
16. Press, W.H., Flannery, B.P., Teukolsky, S.A., Vetterling, W.T.: Numerical Recipes in Pascal: the art of scientific computing. Cambridge University Press (1989)
17. Adenso-Diaz, B., Laguna, M.: Fine-Tuning of Algorithms Using Fractional Experimental Designs and Local Search. Operations Research 54(1), 99–114 (2006)
18. Coy, S., Golden, B., Runger, G., Wasil, E.: Using Experimental Design to Find Effective Parameter Settings for Heuristics. Journal of Heuristics 7(1), 77–97 (2001)
19. Park, M.W., Kim, Y.D.: A systematic procedure for setting parameters in simulated annealing algorithms. Computers and Operations Research 25(3) (1998)
20. Parsons, R., Johnson, M.: A Case Study in Experimental Design Applied to Genetic Algorithms with Applications to DNA Sequence Assembly. American Journal of Mathematical and Management Sciences 17(3), 369–396 (1997)
21. Birattari, M.: The Problem of Tuning Metaheuristics. Phd, Université Libre de Bruxelles (2006)
22. Gaertner, D., Clark, K.L.: On Optimal Parameters for Ant Colony Optimization Algorithms. In: Proceedings of the 2005 International Conference on Artificial Intelligence, vol. 1, pp. 83–89. CSREA Press (2005)
23. Ridge, E., Kudenko, D.: Sequential Experiment Designs for Screening and Tuning Parameters of Stochastic Heuristics. In: Workshop on Empirical Methods for the Analysis of Algorithms, Reykjavik, Iceland. pp. 27–34 (2006)

Comparing Variants of MMAS ACO Algorithms on Pseudo-Boolean Functions*

Frank Neumann[1], Dirk Sudholt[2], and Carsten Witt[2]

[1] Max-Planck-Institut für Informatik, Saarbrücken, Germany
[2] LS 2, FB Informatik, Universität Dortmund, Dortmund, Germany
`fne@mpi-inf.mpg.de`,{`sudholt,cw01`}`@ls2.cs.uni-dortmund.de`

Abstract. Recently, the first rigorous runtime analyses of ACO algorithms have been presented. These results concentrate on variants of the MAX-MIN ant system by Stützle and Hoos and consider their runtime on simple pseudo-Boolean functions such as OneMax and LeadingOnes. Interestingly, it turns out that a variant called 1-ANT is very sensitive to the choice of the evaporation factor while a recent technical report by Gutjahr and Sebastiani suggests partly opposite results for their variant called MMAS. In this paper, we elaborate on the differences between the two ACO algorithms, generalize the techniques by Gutjahr and Sebastiani and show improved results.

1 Introduction

Randomized search heuristics have been shown to be good problem solvers with various application domains. Two prominent examples belonging to this class of algorithms are Evolutionary Algorithms (EAs) and Ant Colony Optimization (ACO) [1]. Especially ACO algorithms have been shown to be very successful for solving problems from combinatorial optimization. Indeed, the first problem where an ACO algorithm has been applied was the Traveling Salesperson Problem (TSP) [2] which is one of the most studied combinatorial problems in computer science.

In contrast to many successful applications, theory lags far behind the practical evidence of all randomized search heuristics. In particular in the case of ACO algorithms, the analysis of such algorithms with respect to their runtime behavior has been started only recently. The analysis of randomized search heuristics (e. g., [3]) is carried out as in the classical algorithm community and makes use of several strong methods for the analysis of randomized algorithms [4], [5].

Regarding ACO, only convergence results [6] were known until 2006 and analyzing the runtime of ACO algorithms has been pointed out as a challenging task in [7]. First steps into analyzing the runtime of ACO algorithms have been made by Gutjahr [8], and, independently, the first theorems on the runtime of a simple ACO algorithm called 1-ANT have been obtained at the same time by

---

* D. S. and C. W. were supported by the Deutsche Forschungsgemeinschaft as a part of the Collaborative Research Center "Computational Intelligence" (SFB 531).

T. Stützle, M. Birattari, and H.H. Hoos (Eds.): SLS 2007, LNCS 4638, pp. 61–75, 2007.
© Springer-Verlag Berlin Heidelberg 2007

Neumann and Witt [9]. Later on this algorithm has been further investigated for the optimization of some well-known pseudo-Boolean functions [10]. A conclusion from these investigations is that the 1-ANT is very sensitive w. r. t. the choice of the evaporation factor ρ. Increasing the value of ρ only by a small amount may lead to a phase transition and turn an exponential runtime into a polynomial one. In contrast to this, a simple ACO algorithm called MMAS_{bs} has been investigated in a recent report by Gutjahr and Sebastiani [11] where this phase transition does not occur. Gutjahr [12] conjectures that the different behavior of MMAS_{bs} and 1-ANT is due to their slightly different replacement strategies: MMAS_{bs} accepts only strict improvements while 1-ANT accepts also equal-valued solutions. We will however show that the replacement strategies do not explain the existence of the phase transition. Instead, the reason is that the 1-ANT only updates pheromone values when the best-so-far solution is replaced.

This motivates us to study MMAS variants where the pheromone values are updated in each iteration. First, we consider the MMAS algorithm by Gutjahr and Sebastiani [11] and show improved and extended results. In particular, we make use of the method called fitness-based partitions which is well-known from the analysis of evolutionary algorithms. Additionally, we study plateau functions and argue why the replacement strategy of the 1-ANT combined with persistent pheromone updates is more natural. Investigating the function NEEDLE, we show that this can reduce the runtime of the ACO algorithm significantly.

The outline of the paper is as follows. In Section 2, we introduce the algorithms that are subject of our investigations. The behavior of these algorithms on well-known plateau functions is considered in Section 3, and Section 4 deals with analyses for some popular unimodal functions. We finish with some conclusions.

2   Algorithms

We consider the runtime behavior of two ACO algorithms. Solutions for a given problem, in this paper bit strings $x \in \{0,1\}^n$ for pseudo-boolean functions $f: \{0,1\}^n \to \mathbb{R}$, are constructed by a random walk on a so-called construction graph $C = (V, E)$, which is a directed graph with a designated start vertex $s \in V$ and pheromone values $\tau: E \to \mathbb{R}$ on the edges.

Algorithm 1 (Construct(C, τ))
1.) $v := s$, mark v as visited.
2.) Let N_v be the set of non-visited successors of v in C. If $N_v \neq \emptyset$:
   a.) Choose successor $w \in N_v$ with probability $\tau_{(v,w)} / \sum_{(v,u)|u \in N_v} \tau_{(v,u)}$.
   b.) Mark w as visited, set $v := w$ and go to 2.).
3.) Return the solution x and the path $P(x)$ constructed by this procedure.

We examine the construction graph given in Figure 1, which is known in the literature as *Chain* [13]. For bit strings of length n, the graph has $3n + 1$ vertices and $4n$ edges. The decision whether a bit x_i, $1 \leq i \leq n$, is set to 1 is made at node $v_{3(i-1)}$. If edge $(v_{3(i-1)}, v_{3(i-1)+1})$ (called 1-edge) is chosen, x_i is set to 1 in the constructed solution. Otherwise the corresponding 0-edge is taken, and $x_i = 0$

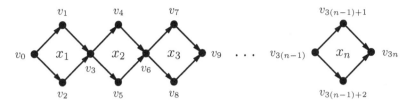

Fig. 1. Construction graph for pseudo-Boolean optimization

holds. After this decision has been made, there is only one single edge which can be traversed in the next step. In case that $(v_{3(i-1)}, v_{3(i-1)+1})$ has been chosen, the next edge is $(v_{3(i-1)+1}, v_{3i})$, and otherwise the edge $(v_{3(i-1)+2}, v_{3i})$ will be traversed. Hence, these edges have no influence on the constructed solution and we can assume $\tau_{(v_{3(i-1)}, v_{3(i-1)+1})} = \tau_{(v_{3(i-1)+1}, v_{3i})}$ and $\tau_{(v_{3(i-1)}, v_{3(i-1)+2})} = \tau_{(v_{3(i-1)+2}, v_{3i})}$ for $1 \leq i \leq n$. We ensure that $\sum_{(u,\cdot) \in E} \tau_{(u,\cdot)} = 1$ for $u = v_{3i}$, $0 \leq i \leq n-1$, and $\sum_{(\cdot,v)} \tau_{(\cdot,v)} = 1$ for $v = v_{3i}$, $1 \leq i \leq n$. Let $p_i = \text{Prob}(x_i = 1)$ be the probability of setting the bit x_i to 1 in the next constructed solution. Due to our setting $p_i = \tau_{(3(i-1),3(i-1)+1)}$ and $1 - p_i = \tau_{(3(i-1),3(i-1)+2)}$ holds, i.e., the pheromone values correspond directly to the probabilities for choosing the bits in the constructed solution. In addition, following the MAX-MIN ant system by Stützle and Hoos [14], we restrict each $\tau_{(u,v)}$ to the interval $[1/n, 1 - 1/n]$ such that every solution always has a positive probability of being chosen.

Depending on whether edge (u, v) is contained in the path $P(x)$ of the constructed solution x, the pheromone values are updated to τ' in the update procedure as follows:

$$\tau'_{(u,v)} = \begin{cases} \min\{(1 - \rho) \cdot \tau_{(u,v)} + \rho, \ 1 - 1/n\} & \text{if } (u, v) \in P(x) \\ \max\{(1 - \rho) \cdot \tau_{(u,v)}, \quad 1/n\} & \text{otherwise.} \end{cases}$$

The following algorithm, which we call MMAS*, has been defined by Gutjahr and Sebastiani [11] under the original name MMAS_{bs}. Here, in each iteration the best solution obtained during the run of the algorithm, called best-so-far solution, is rewarded. Another property of the model is that the best-so-far solution may not switch to another one of the same fitness.

Algorithm 2 (MMAS*)
1.) Set $\tau_{(u,v)} = 1/2$ for all $(u, v) \in A$.
2.) Compute a solution x using $\text{Construct}(C, \tau)$.
3.) Update the pheromone values and set $x^* := x$.
4.) Compute a solution x using $\text{Construct}(C, \tau)$.
5.) If $f(x) > f(x^*)$, set $x^* := x$
6.) Update the pheromone values with respect to x^*.
7.) Go to 4.).

Using this model, it is much easier to adapt results from the well-known evolutionary algorithm called (1+1) EA than in the case of the 1-ANT, i.e., the

MAX-MIN ACO variant examined by Doerr, Neumann, Sudholt and Witt [9], [10]. In particular, many results using the technique of fitness-based partitions may be transferred to MMAS* by taking into account an additional amount of time until the best solution has been rewarded such that an improvement is obtained with a large enough probability (see below for details). This is only possible since pheromone updates occur in each iteration. In contrast to this, the 1-ANT only updates the pheromone values if the best-so-far solution is replaced by solutions of at least the same fitness, i.e., steps 5.) and 6.) in the above description are substituted by

If $f(x) \geq f(x^*)$, set $x^* := x$ and update the pheromone values w.r.t. x^*.

This update strategy may lead to a large discrepancy between the expected value of the next solution and the currently best one and is the main reason for exponential runtimes of the 1-ANT even for really simple functions and relatively large values of ρ (see [9], [10]).

If the value of ρ is chosen large enough in MMAS*, the pheromone borders $1/n$ or $1-1/n$ are touched for every bit of the rewarded solution. In this case, MMAS* equals the algorithm called (1+1) EA*, which is known from the analysis of evolutionary algorithms [15].

Algorithm 3 ((1+1) EA*)
1. *Choose an initial solution $x^* \in \{0,1\}^n$ uniformly at random.*
2. *Repeat*
   a) *Create x by flipping each bit of x^* with probability $1/n$.*
   b) *If $f(x) > f(x^*)$, set $x^* := x$.*

As already pointed out in [15], the (1+1) EA* has difficulties with simple plateaus of constant fitness as no search points of the same fitness as the so far best one are accepted. Accepting solutions with equal fitness enables the algorithm to explore plateaus by random walks. Therefore, it seems more natural to replace search points by new solutions that are at least as good. In the case of evolutionary algorithms, this leads to the well-known (1+1) EA which differs from the (1+1) EA* only in step 2.b) of the algorithm. Similarly, we derive MMAS from MMAS* using this acceptance condition. It should be noted that MMAS is *not* just a variant of the 1-ANT with different pheromone values since it still updates pheromones in each iteration.

Algorithm 4 (Acceptance condition for the (1+1) EA and MMAS)
 – *If $f(x) \geq f(x^*)$, set $x^* := x$.*

In the remainder of the paper, we will examine the behavior of MMAS compared to MMAS*. For the analysis of an algorithm, we consider the number of solutions that are constructed by the algorithm until an optimal solution has been obtained for the first time. This is called the *optimization time* of the algorithm and is a well-accepted measure in the analysis of evolutionary algorithms since each point of time corresponds to a fitness evaluation. Often the expectation of this value is considered and called the *expected optimization time*.

Before we derive the first results, it is helpful to introduce the quantity that informally has been mentioned above. Suppose there is a phase such that MMAS or MMAS* never replaces the best-so-far solution x^* in step 5.) of the algorithm. This implies that the best-so-far solution is rewarded again and again until all pheromone values have reached their upper or lower borders corresponding to the setting of the bits in x^*. The advantage is that probabilities of improvements can be estimated more easily as soon as x^* has been "frozen in pheromone" this way. Gutjahr and Sebastiani [11] call the time for this to happen t^* and bound it from above by $-\ln(n-1)/\ln(1-\rho)$. This holds since a pheromone value which is only increased during t steps is at least $\min\{1 - 1/n, 1 - (1 - 1/n)(1 - \rho)^t\}$ after the iterations, pessimistically assuming the worst-case initialization $1/n$ for this value and $1 - 1/n$ for the complementary pheromone value. In the present paper, we use $\ln(1 - \rho) \leq -\rho$ for $0 \leq \rho \leq 1$ and arrive at the handy upper bound

$$t^* \leq \frac{\ln n}{\rho}. \tag{1}$$

3 Plateau Functions

Plateaus are regions in the search space where all search points have the same fitness. Consider a function $f: \{0,1\}^n \rightarrow \mathbb{R}$ and assume that the number of different objective values for that function is D. Then there are at least $2^n/D$ search points with the same objective vector. Often, the number of different objective values for a given function is polynomially bounded. This implies an exponential number of solutions with the same objective value. In the extreme case, we end up with the function NEEDLE where only one single solution has objective value 1 and the remaining ones get value 0. The function is defined as

$$\text{NEEDLE}(x) := \begin{cases} 1 & \text{if } x = x_{\text{OPT}}, \\ 0 & \text{otherwise}, \end{cases}$$

where x_{OPT} is the unique global optimum. Gutjahr and Sebastiani [11] compare MMAS* and (1+1) EA* w.r.t. their runtime behavior. For suitable values of ρ that are exponentially small in n, the MMAS* has expected optimization time $O(c^n)$, $c \geq 2$ an appropriate constant, and beats the (1+1) EA*. The reason is that MMAS* behaves nearly as random search on the search space while the initial solution of the (1+1) EA* has Hamming distance n to the optimal one with probability 2^{-n}. To obtain from such a solution an optimal one, all n bits have to flip, which has expected waiting time n^n, leading in summary to an expected optimization time $\Omega((n/2)^n)$. In the following, we show a similar result for MMAS* if ρ decreases only polynomially with the problem dimension n.

Theorem 1. *Choosing* $\rho = 1/\text{poly}(n)$, *the optimization time of MMAS* on* NEEDLE *is at least* $(n/6)^n$ *with probability* $1 - e^{-\Omega(n)}$.

Proof. Let x be the first solution constructed by MMAS* and denote by x_{OPT} the optimal one. As it is chosen uniformly at random from the search space, the expected number of positions where x and x_{OPT} differ is $n/2$ and there are at least $n/3$ such positions with probability $1 - e^{-\Omega(n)}$ using Chernoff bounds. At these positions the values of x are rewarded as long as the optimal solution has not been obtained. This implies that the probability to obtain the optimal solution in the next step is at most $2^{-n/3}$. After at most $t^* \leq (\ln n)/\rho$ (see Inequality (1)) iterations, the pheromone values of x have touched their borders provided x_{OPT} has not been obtained. The probability of having obtained x_{OPT} within a phase of t^* steps is at most $t^* \cdot 2^{-n/3} = e^{-\Omega(n)}$. Hence, the probability to produce a solution that touches its pheromone borders and differs from x_{OPT} in at least $n/3$ positions before producing x_{OPT} is $1 - e^{-\Omega(n)}$. In this case, the expected number of steps to produce x_{OPT} is $(n/3)^n$ and the probability of having reached this goal within $(n/6)^n$ steps is at most 2^{-n}.   □

The probability to choose an initial solution x that differs from x_{OPT} by n positions is 2^{-n}, and in this case, after all n bits have reached their corresponding pheromone borders, the probability to create x_{OPT} equals n^{-n}. Using the ideas of Theorem 1 the following corollary can be proved which asymptotically matches the lower bound for the (1+1) EA* given in [11].

Corollary 1. *Choosing $\rho = 1/\mathrm{poly}(n)$, the expected optimization time of MMAS* on* NEEDLE *is $\Omega((n/2)^n)$.*

It is well known that the (1+1) EA that accepts each new solution has expected optimization time $2^{n+o(n)}$ on NEEDLE (see [16], [17]) even though it samples with high probability in the Hamming neighborhood of the latest solution. On the other hand, MMAS* will have a much larger optimization time unless ρ is superpolynomially small (Theorem 1). In the following, we will show that MMAS is competitive with the (1+1) EA even for large ρ-values.

Theorem 2. *Choosing $\rho = \Omega(1)$, the expected optimization time of MMAS on* NEEDLE *is $2^{n+o(n)}$.*

Proof. By the symmetry of the construction procedure and uniform initialization, we w. l. o. g. assume that the needle x_{OPT} equals the all-ones string 1^n. As in [17], we study the process on the constant function $f(x) = 0$. The first hitting times for the needle are the same on NEEDLE and the constant function while the invariant limit distribution for the constant function is easier to study since it is uniform over the search space.

The proof idea is to study a kind of "mixing time" $t(n)$ after which each bit is independently set to 1 with a probability of at least $1/2 - 1/n$ regardless of the initial pheromone value on its 1-edge. This implies that the probability of creating the needle is at least $(1/2-1/n)^n \geq e^{-3}2^{-n}$ (for n large enough) in some step after at most $t(n)$ iterations. We successively consider independent phases of length $t(n)$ until the needle is sampled. Estimating by a geometrically distributed waiting time, the expected optimization time is bounded by $O(t(n) \cdot 2^n)$. The theorem follows if we can show that $t(n) = \mathrm{poly}(n)$.

To bound $t(n)$, we note that the limit distribution of each pheromone value is symmetric with expectation $1/2$, hence, each bit is set to 1 with probability $1/2$ in the limit. We consider a coupling of two independent copies of the Markov chain for the pheromone values of a bit such that we start the chain from two different states. The task is to bound the coupling time $c(n)$, defined as the first point of time where both chains meet in the same border state $1/n$ or $1 - 1/n$ (taking a supremum over any two different initial states). If we know that $E(c(n))$ is finite, then Markov's inequality yields that $c(n) \leq nE(c(n))$ with probability at least $1 - 1/n$, and the coupling lemma [5] implies that the total variation distance to the limit distribution is at most $1/n$ at time $nE(c(n))$. Hence, the bit is set to 1 with probability at least $1/2 - 1/n$ then and we can use $t(n) := nE(c(n))$.

It remains to prove $E(c(n)) = \text{poly}(n)$ for the considered coupling. Since $\rho = \Omega(1)$, the pheromone value of a bit reaches one of its borders in $t^* = O(\log n)$ (Inequality (1)) iterations with probability at least $\prod_{t=0}^{t^*}(1 - (1 - 1/n)(1 - \rho)^t) = \Omega(1/n)$ (using the same estimations for pheromone values as in [11]). Hence, with probability $1/2$, one chain reaches a border in $O(n \log n)$ iterations. A pheromone value that is at a border does not change within $O(n \log n)$ iterations with probability at least $(1 - 1/n)^{O(n \log n)} = 1/\text{poly}(n)$. In this phase, the pheromone value of the other chain reaches this border with probability $\Omega(1)$. Repeating independent phases, we have bounded $E(c(n))$ by a polynomial. $\square$

The function NEEDLE requires an exponential optimization time for each algorithm that has been considered. Often plateaus are much smaller, and randomized search heuristics have a good chance to leave them within a polynomial number of steps. Gutjahr and Sebastiani [11] consider the function NH-ONEMAX that consists of the NEEDLE-function on $k = \log n$ bits and the function ONE-MAX on $n - k$ bits, which can only be optimized if the needle has been found on the needle part. The function is defined as

$$\text{NH-ONEMAX}(x) = \left(\prod_{i=1}^{k} x_i \right) \left(\sum_{i=k+1}^{n} x_i \right).$$

Using the ideas in the proof of Theorem 1 and taking into account the logarithmic size of the needle of NH-ONEMAX, MMAS* with polylogarithmically small ρ finds the needle only after an expected superpolynomial number $2^{\Omega(\log^2 n)}$ of steps, and the following theorem can be shown.

Theorem 3. *Choosing $\rho = 1/\text{polylog}(n)$, the expected optimization time of MMAS* on NH-ONEMAX is $2^{\Omega(\log^2 n)}$.*

As the needle part only consists of $\log n$ bits, MMAS can find the needle after an expected polynomial number of steps (Theorem 2). After this goal has been achieved, the unimodal function ONEMAX has to be optimized. Together with our investigations for unimodal functions carried out in the next section (in particular the upper bound from Theorem 6), the following result can be proved.

Theorem 4. *Choosing $\rho = \Omega(1)$, the expected optimization time of MMAS on NH-ONEMAX is polynomial.*

4   Unimodal Functions, OneMax and LeadingOnes

Gutjahr and Sebastiani [11] extend the well-known fitness-level method, also called the method of f-based partitions, from the analysis of evolutionary algorithms (see, e. g., [18]) to their considered MMAS-algorithm, i. e., the MMAS*. Let $A_1, \ldots, A_m$ be an f-based partition w. r. t. the pseudo-Boolean fitness function $f \colon \{0,1\}^n \to \mathbb{R}$, i. e., for any pair of search points $x \in A_i, y \in A_j$ where $j > i$ it holds $f(x) < f(y)$, and A_m contains only optimal search points. Moreover, let $s_i, 1 \leq i \leq m - 1$, be a lower bound on the probability of the (1+1) EA (or, in this case equivalently, the (1+1) EA*) leaving set A_i. Using the quantity t^*, the expected runtime of MMAS* on f is bounded from above by

$$\sum_{i=1}^{m-1} \left(t^* + \frac{1}{s(i)} \right)$$

(which is a special case of Eq. (13) in [11]). Since $t^* \leq (\ln n)/\rho$, we obtain the more concrete bound

$$\frac{m \ln n}{\rho} + \sum_{i=1}^{m-1} \frac{1}{s(i)}, \tag{2}$$

in which the right-hand sum is exactly the upper bound obtained w. r. t. the (1+1) EA and (1+1) EA*. To prove the bound (2) for MMAS*, it is essential that equally good solutions are rejected (see [11] for a formal derivation). Informally speaking: for each fitness-level, MMAS* in the worst case has to wait until all pheromone values have their obtained upper and lower borders such that the best-so-far solution is "frozen in pheromone" and the situation is like in the (1+1) EA* with the best-so-far solution as the current search point.

In the technical report by Gutjahr and Sebastiani [11], the proposed fitness-level method is basically applied in the context of the unimodal functions ONE-MAX and LEADINGONES. Our aim is to show how the method can be applied to arbitrary unimodal functions both w. r. t. MMAS and MMAS*. Moreover, we generalize the upper bounds obtained in [11] for the example functions ONE-MAX and LEADINGONES and, for the first time, we show a lower bound on the expected optimization time of MMAS* on LEADINGONES. This allows us to conclude that the fitness-level method can provide almost tight upper bounds.

4.1   General Results

Unimodal functions are a well-studied class of fitness functions in the literature on evolutionary computation (e. g., [3]). For the sake of completeness, we repeat the definition of unimodality for pseudo-Boolean fitness functions.

Definition 1. *A function $f \colon \{0,1\}^n \to \mathbb{R}$ is called unimodal if there exists for each non-optimal search point x a Hamming neighbor x' where $f(x') > f(x)$.*

Unimodal functions are often believed to be easy to optimize. This holds if the set of different fitness values is not too large. In the following, we consider unimodal functions attaining D different fitness values. Such a function is optimized by the (1+1) EA and (1+1) EA* in expected time $O(nD)$. This bound is transferred to MMAS* by the following theorem.

Theorem 5. *The expected optimization time of MMAS* on a unimodal function attaining D different fitness values is $O((n + \log n/\rho)D)$.*

Proof. By the introductory argument, we only have to set up an appropriate fitness-based partition. We choose the D sets of preimages of different fitness values. By the unimodality, there is for each current search point x a better Hamming neighbor x' of x in a higher fitness-level set. The probability of the (1+1) EA (or, equivalently, MMAS* with all pheromone values at a border) to produce x' in the next step is $\Omega(1/n)$. By (2), this completes the proof. □

MMAS differs from MMAS* by accepting solutions that are at least as good as the best solution obtained during the optimization process. Hence, the best-so-far solution may switch among several solutions with the same fitness value, and, on every fitness-level, pheromone values may perform random walks between the upper and lower pheromone borders. In comparison to MMAS*, this behavior makes it harder to bound the expected time for an improvement, and the following upper bound is worse than the upper bound for MMAS*.

Theorem 6. *The expected optimization time of MMAS on a unimodal function attaining D different fitness values is $O((n^2(\log n)/\rho)D)$.*

Proof. We only need to show that the expected time for an improvement is $O(n^2(\log n)/\rho)$. The probability that MMAS produces within $O((\log n)/\rho)$ steps a solution being at least as good as (not necessarily better than) the best-so-far solution x^* is $\Omega(1)$ since after at most $(\ln n)/\rho$ steps without exchanging x^* all pheromone values have touched their bounds and then the probability of rediscovering x^* is $\Omega(1)$. We now show that the conditional probability of an improvement if x^* is replaced is $\Omega(1/n^2)$.

Let $x_1, \ldots, x_m$ be an enumeration of all solutions with fitness values equal to the best-so-far fitness value. Due to unimodality, each x_i, $1 \leq i \leq m$, has some better Hamming neighbor y_i; however, the y_i need not be disjoint. Let X and Y denote the event to generate some x_i or some y_i, resp., in the next step. In the worst case $y_1, \ldots, y_m$ are the only possible improvements, hence the theorem follows if we can show $\text{Prob}(Y \mid X \cup Y) \geq 1/n^2$ which is equivalent to $\text{Prob}(Y) \geq \text{Prob}(X)/(n^2 - 1)$.

If $p(x_i)$ is the probability to construct x_i, we have $p(x_i)/p(y_i) \leq (1 - \frac{1}{n})/\frac{1}{n} = n - 1$ as the constructions only differ in one bit. Each y_i may appear up to n times in the sequence $y_1, \ldots, y_m$, hence $\text{Prob}(Y) \geq \frac{1}{n}\sum_{i=1}^{m}\text{Prob}(y_i)$ and

$$\text{Prob}(X) = \sum_{i=1}^{m} p(x_i) \leq (n-1) \cdot \sum_{i=1}^{m} p(y_i) \leq n(n-1) \cdot \text{Prob}(Y).$$

Therefore, $\text{Prob}(Y) \geq \text{Prob}(X)/(n^2 - 1)$ follows. □

Theorems 5 and 6 show that the expected optimization times of both MMAS and MMAS* are polynomial for all unimodal functions as long as $D = \text{poly}(n)$ and $\rho = 1/\text{poly}(n)$. Since MMAS and 1-ANT do not differ in their replacement strategy, this disproves the conjecture in [12] that accepting equally good solutions leads to the phase transition behavior from polynomial to exponential runtimes of the 1-ANT on the following example functions.

4.2 OneMax

The probably most often studied example function in the literature on evolutionary computation is the unimodal function $\text{ONEMAX}(x) = x_1 + \cdots + x_n$. In the runtime analysis of the 1-ANT on ONEMAX by Neumann and Witt [9], it is shown that there exists a threshold value for ρ (in our notation basically $\rho = O(1/n^\epsilon)$ for some small constant $\epsilon > 0$) below which no polynomial runtime is possible. As argued above, due to Theorems 5 and 6, this phase transition can occur neither with MMAS* nor with MMAS. We are interested in improved upper bounds for the special case of MMAS* on ONEMAX. The following theorem has already been proven for some values of ρ in [11]. We however obtain polynomial upper bounds for all ρ bounded by some inverse polynomial.

Theorem 7. *The expected optimization time of MMAS* on* ONEMAX *is bounded from above by* $O((n \log n)/\rho)$.

Proof. The proof is an application of the above-described fitness-level method with respect to the partition $A_i = \{x \mid f(x) = i\}$, $0 \le i \le n$. Using the arguments in [11]—or, equivalently, the upper bound on the expected runtime of the (1+1) EA on ONEMAX [3]—we obtain that the term $\sum_{i=0}^{m-1} 1/(s(i))$ is $O(n \log n)$. Using (2), the upper bound $O((n \log n)/\rho)$ follows. □

The bound is never better than $\Theta(n \log n)$, which is the expected runtime of the (1+1) EA and (1+1) EA* on ONEMAX. At the moment, we are not able to show a matching lower bound $\Omega(n \log n)$ on the expected optimization time of MMAS*; however, we can show that the expected optimization time is growing with respect to $1/\rho$ as the upper bound suggests. We state our result in a more general framework: as known from the considerations by Droste, Jansen and Wegener [3], the mutation probability $1/n$ of the (1+1) EA is optimal for many functions including ONEMAX. One argument is that the probability mass has to be quite concentrated around the best-so-far solution to allow the (1+1) EA to rediscover the last accepted solution with good probability. Given a mutation probability of $\alpha(n)$, this probability of rediscovery equals $(1 - \alpha(n))^n$, which converges to zero unless $\alpha(n) = O(1/n)$. The following lemma exploits the last observation for a general lower bound on the expected optimization time of both MMAS and MMAS*.

Theorem 8. *Let* $f: \{0,1\}^n \to \mathbb{R}$ *be a function with a unique global optimum. Choosing* $\rho = 1/\text{poly}(n)$, *the expected optimization time of MMAS and MMAS* on* f *is* $\Omega((\log n)/\rho)$.

Proof. W. l. o. g., 1^n is the unique optimum. If, for each bit, the success probability (defined as the probability of creating a one) is bounded from above by $1 - 1/\sqrt{n}$ then the solution 1^n is created with only exponentially small probability $(1 - 1/\sqrt{n})^n \leq e^{-\sqrt{n}}$. Using the uniform initialization and pheromone update formula of MMAS and MMAS*, the success probability of a bit after t steps is bounded from above by $1 - (1 - \rho)^t/2$. Hence, all success probabilities are bounded as desired within $t = (\ln(n/4))/(2\rho)$ steps since

$$1 - \frac{1}{2}(1 - \rho)^t \leq 1 - \frac{e^{-(\ln n - \ln 4)/2}}{2} = 1 - \frac{1}{\sqrt{n}}.$$

Since $\rho = 1/\text{poly}(n)$ and, therefore $t = \text{poly}(n)$, the total probability of creating the optimum in t steps is still at most $te^{-\sqrt{n}} = e^{-\Omega(\sqrt{n})}$, implying the lower bound on the expected optimization time. □

Hence, the expected optimization time of MMAS* on ONEMAX is bounded by $\Omega((\log n)/\rho)$, too. It is an open problem to show matching upper and lower bounds. We conjecture that the lower bound for ONEMAX is far from optimal and that $\Omega(n/\rho + n \log n)$ holds.

4.3   LeadingOnes

Another prominent unimodal example function, proposed by Rudolph [19], is

$$\text{LEADINGONES}(x) = \sum_{i=1}^{n} \prod_{j=1}^{i} x_j,$$

whose function value equals the number of leading ones in the considered bit string x. A non-optimal solution may always be improved by appending a single one to the leading ones. LEADINGONES differs from ONEMAX in the essential way that the assignment of the bits after the leading ones do not contribute to the function value. This implies that bits at the beginning of the bit string have a stronger influence on the function value than bits at the end. Because of this, the methods developed by Neumann and Witt [9] cannot be used for analyzing the 1-ANT on LEADINGONES as these methods make particular use of the fact that all bits contribute equally to the function value. In a follow-up paper by Doerr, Neumann, Sudholt and Witt [10], the 1-ANT is studied on LEADINGONES by different techniques and it is shown that a similar phase transition behavior as on ONEMAX exists: for $\rho = o(1/\log n)$ (again using the notation of the present paper), the expected optimization time of the 1-ANT is superpolynomially large whereas it is polynomial for $\rho = \Omega(1/\log n)$ and even only $O(n^2)$ for $\rho = \Omega(1)$. We already know that this phase transition cannot occur with MMAS* and MMAS on LEADINGONES. The following theorem, special cases of which are contained in [11], shows a specific upper bound for MMAS*.

Theorem 9. *The expected optimization time of MMAS* on* LEADINGONES *is bounded from above by* $O((n \log n)/\rho + n^2)$.

Proof. The theorem follows again by the bound (2). We use the same fitness-based partition as in the proof of Theorem 7 and the expected optimization time $O(n^2)$ of the (1+1) EA on LEADINGONES [3]. □

It is interesting that an almost tight lower bound can be derived. The following theorem shows that the expected optimization time of MMAS* is never better than $\Omega(n^2)$. The proof is lengthy, however, for the case of large ρ, one essential idea is easy to grasp: already in the early stages of the optimization process, many, more precisely $\Omega(n)$, pheromone values on 1-edges reach their lower borders $1/n$, and the corresponding bits are set to 0. To "flip" such a bit, events of probability $1/n$ are necessary. This can be transformed into the lower bound $\Omega(n^2)$ on the expected optimization time.

Theorem 10. *Choosing $\rho = 1/\mathrm{poly}(n)$, the expected optimization time of MMAS* on LEADINGONES is bounded from below by $\Omega(n/\rho + n^2)$.*

Proof. We first show the lower bound $\Omega(n/\rho)$ on the expected optimization time. Afterwards, this bound is used to prove the second bound $\Omega(n^2)$. Throughout the proof, we consider only runs of polynomial length since by the assumption $\rho = 1/\mathrm{poly}(n)$ the lower bounds to show are both polynomial. This allows us to ignore events with exponentially small probabilities.

For the first part, we use the observation by Doerr et al. [10] on the pheromone values outside the block of leadings ones. If the best-so-far LEADINGONES-value equals k, the pheromone values corresponding to the bits $k+2, \ldots, n$ have never contributed to the LO-value implying that each of these bits is unbiasedly set to 1 with probability exactly $1/2$ in the next constructed solution. Note that these pheromone values are not necessarily martingales, however, the probability distribution of such a pheromone value is symmetric with expectation $1/2$. Hence we distinguish two states for a bit: if it is right of the leftmost zero in the best-so-far solution, its expected pheromone values on 1- and 0-edges equal $1/2$; if it is left of the leftmost zero, the pheromone value of its 1-edges are monotonically increasing in each iteration until the border $1 - 1/n$ is reached. We call the first state the *random state* and the second one the *increasing state*.

While all bits in the block $n/4 + 1, \ldots, n/2$ are in random state, the probability of obtaining a LEADINGONES-value of at least $n/2$ is bounded from above by $2^{-\Omega(n)}$. Hence, with probability $1 - 2^{-\Omega(n)}$, the LEADINGONES-value is in the interval $[n/4, n/2]$ at some point of time. The randomness of the bits after the leftmost zero allows us to apply the standard free-rider arguments by Droste, Jansen and Wegener [3]. Hence, with probability $1 - 2^{-\Omega(n)}$, at least $n/12$ improvements of the LEADINGONES-value have occurred when it has entered the interval $[n/4+1, n/2]$. This already completes the first part of the proof if $\rho = \Omega(1)$. In the following, we therefore study the case $\rho = o(1)$. Assume for some arbitrarily slowly increasing function $\alpha(n) = \omega(1)$ that only $n/(\alpha(n)^2\rho)$ iterations have happened until the first $n/12$ improvements are done. Then it must hold (by the pigeon-hole principle) that at most $n/\alpha(n)$ times, the number of iterations between two consecutive improvements is large, which is defined as: *not* bounded by $O(1/(\alpha(n)\rho))$. Furthermore, by another pigeon-hole-principle argument, there must be at least $n/\alpha(n)$ independent so-called fast phases defined as follows: each fast phase consists of at least $r := \lfloor \alpha(n)/24 \rfloor$ consecutive improvements between which the number of iterations is never large, i.e., each time bounded by $O(1/(\alpha(n)\rho))$. (For a proof, consider a fixed subdivision of $n/12$ improvements into blocks of r consecutive improvements and show that at most half of these blocks can contain a large number of iterations between some two consecutive improvements.) In the following, we will show that a fast phase has probability $o(1)$. This contradicts (up to failure probabilities of $2^{-\Omega(n/\alpha(n))}$) the assumption we have at least $\Omega(n/\alpha(n))$ fast phases, hence the number of iterations for the first $n/12$ improvements cannot be bounded by $n/(\alpha(n)^2\rho)$. Since $\alpha(n)$ can be made arbitrarily slowly increasing, we obtain the lower bound $\Omega(n/\rho)$ on the expected optimization time.

Consider the event that a fast phase containing r improvements is sufficient to set at least r bits into the increasing state (the phase is called successful then). In the beginning, all these bits are in random state. Hence, with a failure probability at most $2^{-\Omega(r)}$, less than $r/4$ pheromone values on the corresponding 1-edges (in the following called success probabilities) are at most $1/2$. We assume $r/4$ bits with this property and estimate the probability of setting all these bits to 1 simultaneously in at least one improving step until the end of the phase (which is necessary for the phase to be successful). The success probability of a bit with initial pheromone value $1/2$ is still at most $p_t := 1 - (1-\rho)^t/2$ if it has been only in increasing state for t steps. The total number of iterations in the phase is $O(1/\rho)$ by the definition of a fast phase. Hence, by the end of the phase, all considered success probabilities are at most $1 - (1-\rho)^{O(1/\rho)}/2 = 1 - \Omega(1)$. The probability of a single improving step setting the $r/4$ bits to 1 is therefore at most $(1 - \Omega(1))^{r/4} = 2^{-\Omega(r)}$. Adding up over all r improving steps and taking into account the above failure probability, the probability of the phase being successful is at most $r2^{-\Omega(r)} + 2^{-\Omega(r)} = 2^{-\Omega(\alpha(n))} = o(1)$ as suggested.

Having proved that the expected number of steps for $n/12$ improvements is $\Omega(n/\rho)$, we conclude that there is a constant $c > 0$ such that the time between two improvements is at least c/ρ with probability at least $1/2$. Otherwise Chernoff bounds would (up to exponentially small failure probabilities) contradict the bound $\Omega(n/\rho)$. We exploit this to show that with high probability, a linear number of bits in random state reaches the lower border $1/n$ on the success probability during the optimization process. This will prove the second lower bound $\Omega(n^2)$ on the expected optimization time.

Consider a bit in random state and a phase of $t^* \leq (\ln n)/\rho$ iterations. We are interested in the event that the bit is set to zero throughout all improvements of the phase, which implies that the success probability is $1/n$ until the end of the phase (the phase is finished prematurely if the border $1/n$ is reached in less than t^* steps). This event has probability $\Omega(1)$ for the following reasons: let p_0 be the initial probability of setting the bit to zero and assume $p_0 \geq 1/2$ (which holds with probability at least $1/2$). After t steps of the phase, this probability is at least $p_t := 1 - (1-\rho)^t/2$ if the bit has only been set to zero in the improvements in the phase. If the time between two improvements (and, therefore, exchanges of the best-so-far solution) is always bounded from below by c/ρ, the considered event has probability at least $\prod_{t=1}^{\infty} p_{tc/\rho}$. Using $1 - x \leq e^{-x}$ for $x \in \mathbb{R}$ and $1 - x \geq e^{-2x}$ for $x \leq 1/2$, this term can be bounded from below by

$$\prod_{t=1}^{\infty} \left(1 - \frac{(1-\rho)^{tc/\rho}}{2}\right) \geq \prod_{t=1}^{\infty} \left(1 - \frac{e^{-ct}}{2}\right) \geq \prod_{t=1}^{\infty} \exp(e^{-ct})$$

$$= \exp\left(\sum_{t=1}^{\infty} e^{-ct}\right) = \exp\left(\frac{e^{-c}}{1 - e^{-c}}\right) = \Omega(1).$$

That the phase is of the desired length only with probability $1/2$ is no problem either since we can bound the probability of setting the bit to 0 after a short phase by the probability in the beginning of the phase and argue like in the proof of Theorem 2 in [10]. The event of a short phase corresponds to the occurrence of a "free-rider" and the number of short phases is geometrically distributed with success probability at most $1/2$. Hence, using the same calculations as in the mentioned proof, the probability of the desired event is at least $\prod_{t=1}^{\infty} \frac{p_{tc/\rho}}{2 - p_{tc/\rho}} \geq$ $\prod_{t=1}^{\infty} \frac{p_{tc/\rho}}{2} = \Omega(1)$ as claimed.

Using the observation from the last paragraph, it follows that with probability $1 - 2^{-\Omega(n)}$, at least $\Omega(n)$ bits in random state have reached success probability $1/n$ by the time the LEADINGONES-value enters the interval $[n/4, n/2]$. The only problem might be that these success probabilities might increase again. However, in each of the remaining improvements, we distinguish the events whether it is relevant for an improvement to set such a bit to 1 or not. If the bit is not relevant since it is still right of the leftmost zero after the improvement, its success probability does not change with probability $1 - 1/n$. Hence, in $O(n)$ improvements there are in expectation still $\Omega(n)$ success probabilities equal to $1/n$ left. Since, with at least constant probability, $\Omega(n)$ improvements are necessary, the lower bound $\Omega(n^2)$ on the expected optimization time follows.   □

In the proof, we had to carefully look at the random bits and to study the structure of the optimization process. It seems to be even harder to prove a corresponding lower bound for MMAS since accepting equally good solutions implies that more than n exchanges of the best-so-far solution can happen. Also additional ideas are required to transfer the proof of Theorem 10 and to obtain an improved lower bound for MMAS* on ONEMAX.

5   Conclusions

The rigorous runtime analysis of ACO algorithms is a challenging task where the first results have been obtained only recently. In this paper, we have considered an ACO algorithm called MMAS for which some results based on the method of fitness-based partitions have been obtained. Previous results on this algorithm by Gutjahr and Sebastiani have been extended and improved and compared to our earlier findings for the 1-ANT. In particular, we have considered some unimodal functions such as ONEMAX and LEADINGONES and proved upper and lower bounds. Furthermore, we have argued why it is necessary to replace search points by other ones that have the same fitness and shown that this improves the runtime on the well-known plateau function NEEDLE.

Acknowledgements. The authors thank the reviewers for their helpful comments.

References

1. Dorigo, M., Stützle, T.: Ant Colony Optimization. MIT Press, Cambridge (2004)
2. Dorigo, M., Maniezzo, V., Colorni, A.: The ant system: An autocatalytic optimizing process. Technical Report 91-016 Revised, Politecnico di Milano (1991)
3. Droste, S., Jansen, T., Wegener, I.: On the analysis of the (1+1) evolutionary algorithm. Theor. Comput. Sci. 276, 51–81 (2002)
4. Motwani, R., Raghavan, P.: Randomized Algorithms. Cambr. Univ. Press, Cambridge (1995)
5. Mitzenmacher, M., Upfal, E.: Probability and Computing – Randomized Algorithms and Probabilistic Analysis. Cambr. Univ. Press, Cambridge (2005)
6. Gutjahr, W.J.: ACO algorithms with guaranteed convergence to the optimal solution. Inform. Process. Lett. 82, 145–153 (2002)
7. Dorigo, M., Blum, C.: Ant colony optimization theory: A survey. Theor. Comput. Sci. 344, 243–278 (2005)
8. Gutjahr, W.J.: First steps to the runtime complexity analysis of Ant Colony Optimization. Comput. Oper. Res (to appear)
9. Neumann, F., Witt, C.: Runtime analysis of a simple ant colony optimization algorithm. In: Asano, T. (ed.) ISAAC 2006. LNCS, vol. 4288, pp. 618–627. Springer, Heidelberg (2006) extended version to appear in Algorithmica
10. Doerr, B., Neumann, F., Sudholt, D., Witt, C.: On the runtime analysis of the 1-ANT ACO algorithm. In: Proc. of GECCO '07, ACM Press, New York (to appear)
11. Gutjahr, W.J., Sebastiani, G.: Runtime analysis of ant colony optimization. Technical report, Mathematics department, "Sapienza" Univ. of Rome, 2007/03 (2007)
12. Gutjahr, W.J.: Mathematical runtime analysis of ACO algorithms: Survey on an emerging issue. Swarm Intelligence (to appear)
13. Gutjahr, W.J.: On the finite-time dynamics of ant colony optimization. Methodol. Comput. Appli. Probab. 8, 105–133 (2006)
14. Stützle, T., Hoos, H.H.: MAX-MIN ant system. J. Future Gener. Comput. Syst. 16, 889–914 (2000)
15. Jansen, T., Wegener, I.: Evolutionary algorithms - how to cope with plateaus of constant fitness and when to reject strings of the same fitness. IEEE Trans. Evolut. Comput. 5(6), 589–599 (2001)
16. Garnier, J., Kallel, L., Schoenauer, M.: Rigorous hitting times for binary mutations. Evolut. Comput. 7, 173–203 (1999)
17. Wegener, I., Witt, C.: On the optimization of monotone polynomials by simple randomized search heuristics. Combin. Probab. Comput. 14, 225–247 (2005)
18. Wegener, I.: Methods for the analysis of evolutionary algorithms on pseudo-boolean functions. In: Sarker, R., Yao, X., Mohammadian, M. (eds.) Evolutionary Optimization, pp. 349–369. Kluwer, Dordrecht (2002)
19. Rudolph, G.: Convergence Properties of Evolutionary Algorithms. Kovač (1997)

EasyAnalyzer: An Object-Oriented Framework for the Experimental Analysis of Stochastic Local Search Algorithms

Luca Di Gaspero[1], Andrea Roli[2], and Andrea Schaerf[1]

[1] DIEGM, University of Udine, Udine, Italy
[2] DEIS, University of Bologna, Cesena, Italy
{l.digaspero,schaerf}@uniud.it, andrea.roli@unibo.it

Abstract. One of the aspects of applying software engineering to Stochastic Local Search (SLS) is the principled analysis of the features of the problem instances and the behavior of SLS algorithms, which —because of their stochastic nature— might need sophisticated statistical tools.

In this paper we describe EASYANALYZER, an object-oriented framework for the experimental analysis of SLS algorithms, developed in the C++ language. EASYANALYZER integrates with EASYLOCAL++, a framework for the development of SLS algorithms, in order to provide a unified development and analysis environment. Moreover, the tool has been designed so that it can be easily interfaced also with SLS solvers developed using other languages/tools and/or with command-line executables.

We show an example of the use of EASYANALYZER applied to the analysis of SLS algorithms for the k-GRAPHCOLORING problem.

1 Introduction

In recent years, much research effort has focused on the proposals of environments specifically designed to help the formulation and implementation of Stochastic Local Search (SLS) algorithms by means of specification languages and/or software tools, such as LOCALIZER and its evolutions [3,1,2], HOTFRAME [4], ParadisEO [5], iOpt [6], EASYLOCAL++ [7,8], and others.

Unfortunately, as pointed out by Hoos and Stützle [9] in [9, Epilogue, pp. 533–534], the same amount of effort has not been oriented in the development of software tools for the experimental analyses of the algorithms.

To this regard, [10] proposes a suite of tools for visualizing the behavior of SLS algorithms, which is particularly tailored for MDF (Metaheuristics Development Framework) [11]. However, to the best of our knowledge, we can claim that at present there is no widely-accepted comprehensive environment.

In this paper we try to overcome this lack by proposing an object-oriented framework, called EASYANALYZER, for the analysis of SLS algorithms. EASYANALYZER is a software tool that belongs to the family of Object-Oriented (O-O) frameworks. A framework is a special kind of software library, which consists of a hierarchy of abstract classes and is characterized by the *inverse control*

T. Stützle, M. Birattari, and H.H. Hoos (Eds.): SLS 2007, LNCS 4638, pp. 76–90, 2007.

mechanism for the communication with the user code (also known as the *Hollywood Principle*: "Don't call us, we'll call you"). That is, the functions of the framework call the user-defined ones and not the other way round as it usually happens with software libraries. The framework thus provides the full control logic and, in order to use it, the user is required to supply the problem specific details by means of some standardized interfaces.

Our work is founded on *Design Patterns* [12], which are abstract structures of classes, commonly present in O-O applications and frameworks, that have been precisely identified and classified. The use of patterns allows us to address many design and implementation issues in a more principled way.

EASYANALYZER provides a family of off-the-shelf analysis methods to be coupled to local search solvers developed using one of the tools mentioned above or written from scratch. For example, it performs various kinds of search space analysis in order to understand, study, and tune the behavior of SLS algorithms. The properties of the search space are a crucial factor of SLS algorithm performance [13,9]. Such characteristics are usually studied by implementing *ad hoc* programs, tailored both to the specific algorithm and to the problem at hand. EASYANALYZER makes it possible to abstract from algorithm implementation and problem details and to design general search space analyzers.

EASYANALYZER is specifically designed to blend in a natural way with EASYLOCAL++, the local search framework developed by two of these authors [8,7], which has recently been entirely redesigned to allow for more complex search strategies. Nevertheless, it is capable of interacting with other software environments and with stand-alone applications.

This is an ongoing work, and some modules still have to be implemented. However, the general architecture, the core modules, and the interface with EASYLOCAL++ and with command-line executables are completed and stable.

The paper is organized as follows. In Section 2 we show the architecture of EASYANALYZER and its main modules. In Section 3 we go in details in the implementation of the core modules. In Section 4 we show some examples of use based on the classic k-GRAPHCOLORING problem. In Section 5 we draw some conclusions and discuss future work.

2   The Architecture of EasyAnalyzer

The conceptual architecture of EASYANALYZER is presented in Figure 1 and it is split in three main abstraction layers. Each layer of the hierarchy relies on the services supplied by lower levels and provides a set of more abstract operations.

Analysis system: it comprises the *core classes* of EASYANALYZER. It is the most abstract level and contains the control logic of the different types of analysis provided in the system. The code for the analyses is completely abstract from the problem at hand and also from the actual implementation of the solver. The classes of this layer delegate implementation- and/or problem-related tasks to the set of lower level classes, which comply with a predefined service interface (described in the following).

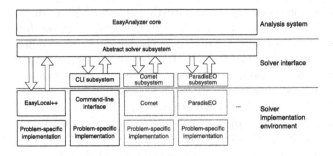

Fig. 1. EASYANALYZER layered architecture

Solver interfaces: this layer can be split into two components: the top one is the interface that represents an abstract solver subsystem, which simply prescribes the set of services that should be provided by a concrete solver in order to be used in the analyses. The coupling of the analysis system with the implementation is dealt with by this component.

The lower component is the concrete implementation of the interface for a set of SLS software development environments. Notice that in the case of EASYLOCAL++, this component is not present since EASYANALYZER directly integrates within the development framework classes. The reason is that in the design of the solver interface we reuse many choices already made for EASYLOCAL++ thus allowing immediate integration.

For other software environments, instead, the solver subsystem component must be explicitly provided. Depending on the capabilities of the software environment, these interfaces can be implemented in a problem-independent manner (so that they can be directly reused across all applications) or it might require to be customized for the specific problem. Although in the second case the user could be required to write some additional code, our design limits this effort since our interfaces requires just a minimal set of functionalities.

Solver environment: it consists of the (possibly generic) SLS software development environment plus the problem-specific implementation. In some cases these two components coincide, as for solvers that do not make use of any software environment. In this case the interaction with the solver can make use of a simple command-line interface.

At present, we have implemented the direct integration with EASYLOCAL++ and to the command-line interface[1] by means of a set of generic classes (i.e., C++ classes that make use of templates that should be instantiated with the concrete command-line options). We plan to implement also the interfaces to other freely available software environments like, e.g., ParadisEO [5] and Comet [2].

---

[1] In Figure 1 the implemented components are denoted by solid lines while dotted lines denote components only designed.

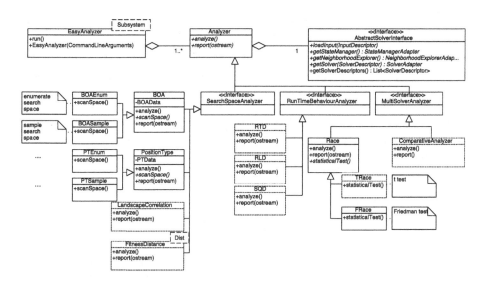

Fig. 2. UML class diagram of the analysis system

In the following subsections we present more in detail the problem-independent layers of the EASYANALYZER architecture and we give some example of code.

2.1   The Analysis System

The main classes of the analysis system are shown in Figure 2 using the UML 2.0 notation [14]. As in Figure 1 we report in solid lines the fully implemented components (dotted lines for the forthcoming ones).

Let us start our presentation with the `EasyAnalyzer` class. This class relies on the *Factory method* pattern to set up the analysis system on the basis of a given solver interface. Notice that the interface is specified as a template parameter, so that we are able to write the generic code for instantiating the analysis system regardless which of the concrete implementations is provided. Furthermore, the `EasyAnalyzer` class provides a standardized command-line interface for the interaction with the analysis system. This task is accomplished by managing a command-line interpreter object that is directly configured by the analysis techniques. That is, each analysis technique "posts" the syntax of the command-line arguments needed by the interpreter object that is in charge of parsing the command line and dispatching the actual parameters to the right component.

The main component of the analysis system is the `Analyzer` class, which relies on the *Strategy* pattern. This component represents the interface of an analysis technique, whose actual "strategy" will be implemented in the `analyze()` method defined in the concrete subclasses. The `report(ostream)` method is used to provide on an output stream a human- and/or machine-readable report of the analysis, depending on the parameters issued on the command line.

The `Analyzer` class is then specialized on the basis of the SLS features that are subject of the analysis into the following three families:

SearchSpaceAnalyzer: these analyzers deal with features that are related to the search space. Several crucial properties of the search space can be analyzed with these modules, such as landscape characteristics and states reachability.

RunTimeBehaviorAnalyzer: their aim is to analyze the run-time behavior of the solvers. Analyses belonging to this family are, e.g., *run-time distribution* (`RTD`), *run-length distribution* (`RLD`) and *solution quality distribution* (`SQD`).

MultiSolverAnalyzer: they handle and evaluate groups of solvers. For example the `Race` analyzer tries to find-out the statistically best configuration of a solver among a set of candidate configurations by applying a racing procedure [15].

The interface with the services provided by the analysis system is established with the `AbstractSolverInterface` abstract class, which relies on the *Façade* pattern whose aim is to provides a simple interface to a complex subsystem. This class and the underlying classes and objects responsibilities are going to be detailed in the following subsection.

2.2   The Solver Interface

The architecture of the solver interface is shown in the top part of Figure 3. The derived classes on the bottom are the implementation of this interface in the EASYLOCAL++ framework.

The `SolverInterface` class acts as a unified entry point (the *Façade*) and as the coordinator of a set of underlying classes (*Abstract Factory* and *Factory method* patterns). Indeed, according to the EASYLOCAL++ design, we identify a set of software components that take care of different responsibilities in a SLS algorithm and we define a set of *adapter* classes for them. These adapters have a straight implementation in EASYLOCAL++ (Figure 3, bottom part), and are those components that instead must be implemented for interfacing with different software environments. The components we consider are the following:

StateManagerAdapter: it is responsible for all operations on the states of the search space that are independent of the definition of the neighborhood. In particular, it provides methods to enumerate and to sample the search space, and it allows us to evaluate the cost function value on a given state. The component relies on `StateDescriptor`s for the exchange of information with the analysis system (in order to avoid the overhead of sending a complex space representation).

NeighborhoodExplorerAdapter: it handles all the features concerning the exploration of the neighborhood. It allows to enumerate and to sample the neighbors of a given state, and to evaluate the cost function.

SolverAdapter: it encapsulates a single SLS algorithm or a complex solution strategy that involves more than one single SLS technique. Its methods allow us to perform a full solution run (either starting from a random initial state

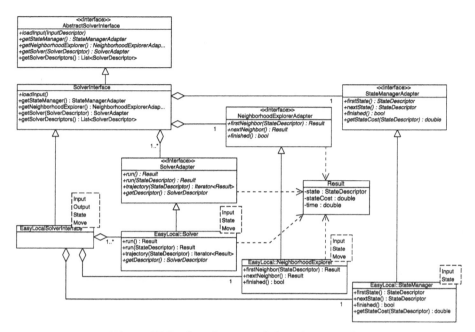

Fig. 3. UML class diagram of the solver interface

or from a state given as input), possibly storing all the trajectory from the initial state to the final one. This component returns also information on the running time and on the state costs.

2.3   How to Use EasyAnalyzer

In order to use EASYANALYZER it is only needed to instantiate the `Solver` template of the `EasyAnalyzer` class with the proper implementation of the `Abstract SolverInterface`. As for the EASYLOCAL++ solver, this interface is already provided with the framework, whilst for the command-line interaction the functionalities must be implemented by the user in the stand-alone executable.

The various analyses can be executed by issuing command line options to the EASYANALYZER executable. For example, `-ptenum` requires a position type analysis to be performed (with a complete enumeration of the search space). Additional parameters, depending on the analysis at hand, can be required and they have to be specified on the command line as well. The different types of analysis and an explanation of the options available can be obtained by issuing a `-help` command.

3   Implementation of EasyAnalyzer

In this section we describe two representative examples of the analyzers currently implemented, with emphasis on the design process that relies on the abstractions provided by the Solver interfaces.

3.1 SearchSpaceAnalyzer

In this section we illustrate the design and implementation of an analyzer for *Basins of attraction* (BOA), useful for studying the reachability of solutions. Given a deterministic algorithm, the basin of attraction $\mathcal{B}(\bar{s})$ of a search space state $\bar{s}$ (usually a minimum), is defined as the set of states that, taken as initial states, give origin to trajectories that end at point $\bar{s}$. The quantity $rBOA(\bar{s})$, defined as the ratio between the cardinality of $\mathcal{B}(\bar{s})$ and the search space size (assumed finite), is an estimation of the reachability of state $\bar{s}$. If the initial solution is chosen at random, the probability of finding a global optimum s^* is exactly equal to $rBOA(s^*)$. Therefore, the higher is this ratio, the higher is the probability of success of the algorithm. The estimation of basins of attraction characteristics can help in the *a posteriori* analysis of local search performance, to provide explanations for the observed behavior. Moreover, it can also be useful for the *a priori* study of the most suitable models of a problem, for instance for comparing advantages and disadvantages of models that incorporate symmetry-breaking or implied constraints [16]. In section 4.2, we will discuss an example of a typical application of this kind of *a posteriori* analysis.

The development of a specific analyzer starts from the implementation of the interface `SearchSpaceAnalyzer` that declares the basic methods `analyze()`, for the actual analysis to be performed, and `report()`, defining the output of the analysis. The main goal of a BOA analyzer is to find the size of all, or a sample of, the local and global minima basins of attraction, corresponding to the execution of a given (deterministic) algorithm $\mathcal{A}$. Therefore, a BOA analyzer must be fed with problem instance and search algorithm and its task is to scan the search space for finding attractors and their basins. The procedure of search space scanning can be implemented in several ways, and it could primarily be either an exhaustive enumeration or a sampling. Attractors and their basins can be then computed by running algorithm $\mathcal{A}$ from every possible initial state s, returned by the scan method, till the corresponding attractor.[2] The main parts of the `analyze()` method for the BOA class are as detailed in Listing 1.1.

Listing 1.1. The `analyze()` method for the BOA class

```
void BOA::analyze()
{ BOAData data;
  initializeAnalysis(); // loads instance and solver
  StateDescriptor state = scanSpace();
  while (state.isValid()) // while there are feasible states
  { const Result& result = solver.run(state);
    updateBOAInfo(result.getStateDescriptor());
    state = scanSpace();
  }
}
```

[2] There are also other ways for performing this task; for instance, the *Reverse hill-climbing* technique [17]. Moreover, in this discussion, we only consider the case of deterministic algorithms.

The BOA analyzer is designed through the *Template Method* pattern, that enables the designer to define a class that delegates the implementation of some methods to the subclasses. In this case, the implementation of the method `scanSpace()` is left to the subclasses, so as to make it possible to implement a variety of different search space scanning procedures, such as enumeration (`BOAEnum`, Listing 1.2) and uniform sampling (`BOASample`, Listing 1.3). These methods rely on `StateManagerAdapter` for enumeration and random sampling of the search space, respectively.

Listing 1.2. `BOAEnum::scanSpace()`

```
StateDescriptor
  BOAEnum::scanSpace()
{
  if(!stateManager.finished())
    return
      stateManager.nextState();
  else return NON_VALID_STATE;
}
```

Listing 1.3. `BOASample::scanSpace()`

```
StateDescriptor
  BOASample::scanSpace()
{ if (numberOfSamples <
     maxNumberOfSamples)
    return
      stateManager.randomState();
  else return NON_VALID_STATE;
}
```

We remark that the implementation of these BOA analyzers is very simple and compact, as it is totally independent from the problem specific part of the software, thanks to the intermediate software level of Solver interfaces. With few lines of code it is possible to implement other BOA analyzers, for instance by using samplings based on non-uniform distributions, in order to bias the sampling in areas containing global minima.

In an analogous way, it is possible to implement analyzers which can scan the search space and classify each state as (strict) local minimum/maximum, plateau, slope or ledge as a function of the cost of its neighbors. According to [9], we call this kind of classification *position type* analysis. The class `PositionType` delegates the subclasses `PTEnum` and `PTSample` for the implementation of the method `scanSpace()`, that relies on the class `NeighborhoodExploreAdapter` for enumerating the neighborhoods. The method `scanSpace()` can enumerate or sample the search space. The current implementation includes enumeration and uniform sampling, while the sampling through different distributions or along trajectories is part of ongoing work.

3.2 MultiSolverAnalyzer

In many cases, the people working on SLS algorithms face the problem of evaluating the behavior of a family of solvers (usually on a set of a benchmark instances) rather than analyzing a single SLS algorithm. For example one could be interested in comparing a set of SLS solvers to determine whether one or more of them perform better than the others. Another common case is to consider different settings for the same solver as a mean for tuning the parameters of the solver. In both cases statistical procedures are needed to assess the choice of the "winning" solver in a sound way.

To deal with this situation, we decided to design also a set of analyzers that manage a set of SLS solver and whose aim is to perform comparative analysis. As in the previous example, we rely on the abstraction levels of EASYANALYZER to design general multi-solver analyzers.

We have developed the set of classes that implement the Race approach by Birattari et al. [15]. This procedure aims at selecting the parameters of a SLS algorithm by testing each candidate configuration on a set of trials. The configurations that perform poorly are discarded and not tested anymore as soon as sufficient statistical evidence against them is collected.

This way, only the statistically proven good configurations continue the race, and the overall number of tests needed to find the best configuration (or the equally good configurations) is limited. Each trial is performed on the same randomly chosen problem instance for all the remaining configurations and a statistical test is used to assess which of them are discarded.

In order to perform the analysis, the user must specify a set of solvers that are going to be compared in the Race and a set of instances on which the solvers will be run.

We present here the method `analyze()` of the class `Race` (Listing 1.4). The method works in a loop that evaluates the behavior of the configurations on an instance and collects statistical evidence about them. We would like to remark that our implementation follows the lines of the R package [18].

Listing 1.4. The `analyze()` method for the class `Race`

```
void Race::analyze()
{ initializeAnalysis(); // loads instances, solvers and sets up the set
    of aliveSolvers
  replicate = 0;
  do
  { performReplicate(instances[replicate % instances.size()], replicate);
    if (replicate >= min_replicates) // the test is performed only after
        a minimum number of replicates
    { TestResult res = statisticalTest(seq(0, replicate), aliveSolvers,
        conf_level);
      updateAliveSolvers(res.survived);
      statistics[replicate] = res.statistic;
      p_values[replicate] = res.p_value;
    }
    replicate++;
  }
  while (aliveSolvers.size() > 1 && replicate < max_replicates);
}
```

The evaluation of the candidate configurations is performed by calling the method `performReplicate` (whose code is reported in Listing 1.5). This method relies on the solver interfaces to load the current input instance and to invoke the different solvers configurations (only for the solvers that still survive the race).

Listing 1.5. The `performReplicate()` method of the class `Race`

```
void Race::performReplicate(const std::string& instance, unsigned int i)
{ sub.loadInput(instance);
  for (unsigned int j = 0; j < solvers.size(); j++)
    if (aliveSolvers.find(j) != aliveSolvers.end())
    { Result r = solvers[j]->run();
      outcomes[i][j] = r.getCostValue();
    }
}
```

The class `Race` makes use of the *Template Method* pattern: the selection algorithm relies on the implementation of the abstract `statisticalTest()` method, which is implemented in two different sub-classes for the Student's t-test (`TRace`) and the Friedman's test (`FRace`).

Notice that the presented method makes use of the solvers just as black-boxes that from an initial state lead to a final solution. Indeed, the only information exploited in the analysis is the final solution cost and the running time. More sophisticated analyses can also exploit the trajectory from the initial state to the solution. For example this information can be used to compare the quality of SLS solvers throughout the evolution of the search (as suggested by Taillard [19]). This will be subject of future work.

4 A Case Study: The k-GraphColoring Problem

We show an example of the use of EASYANALYZER by providing some analyses on a family of solvers for the k-GRAPHCOLORING problem. Our aim is not to say the ultimate word on the problem, but rather to exemplify the use of the analyzers presented so far.

4.1 k-GraphColoring Problem Statement and Local Search Encoding

Here we briefly recall the statement of the k-GRAPHCOLORING problem, which is the decision variant of the well-known min-GRAPHCOLORING problem [20, Prob. GT4, page 191].

Given an undirected graph $G = (V, E)$ and a set of k integer colors, the problem is to assign to each node $v \in V$ a color value $c(v)$ such that adjacent nodes are assigned different colors.

As the search space of our SLS algorithms we consider the set of all possible colorings of the graph, including the infeasible ones; the number of conflicting nodes is the cost function value in a given state. The neighborhood relation

Table 1. Position types in 3-colorable graphs

Position type	Edge Density						
	0.010	0.016	0.020	0.026	0.030	0.036	0.040
Strict local min	$< 10^{-4}$	0%	0%	0%	0%	0%	0%
Local min	0%	0%	0%	0%	0%	0%	0.46%
Interior plateau	0%	0%	0%	0%	0%	0%	92.74%
Ledge	84.42%	99.42%	99.80%	100%	100%	100%	0.10%
Slope	0.77%	0.02%	$< 10^{-4}$	0%	0%	0%	0%
Local max	14.77%	0.56%	0.20%	$< 10^{-4}$	$< 10^{-4}$	0%	6.70%
Strict local max	0.04%	0%	0%	0%	0%	0%	0%

is defined by the color change of one conflicting node (as in [21]) and for the tabu search prohibition mechanism, we consider a move inverse of another one if both moves insist on the same node and the first move tries to restore the color changed by the second one.

4.2   Search Space Analysis

To illustrate the use of EASYANALYZER for studying properties of the search space, we consider a simple analysis on 3-colorable graphs. Instances were generated with Culberson's graph generator [22] with equi-partition and independent random edge assignment options and with varying edge density, so as to span the spectrum from lowly to highly constrained instances. All instances are guaranteed to be 3-colorable and have 100 nodes.

One of the main search space features of interest is the number of local minima and, more generally, the type of search space positions. Table 1 reports a summary of the position type analysis out of 10^6 random samples.

As discussed in [9], for random landscapes we would expect a position type distribution characterized by a majority of least constrained positions, such as ledges, which are states with neighbors with higher, lower and equal cost. From the results in Table 1, we observe that ledge is the predominant type. The most constrained instance (density = 0.04) shows instead a very different landscape structure, as it is dominated by plateaus. This difference with respect to the other instances is particularly apparent also when local search is used to solve these instances. Figure 4 shows the box-plots corresponding to the execution of 100 independent short runs of a simple hill-climbing (draw a random move, accept it if improving or sideways). The algorithm stops after 10 iterations without improvements. The performance of local search on the most constrained instance is significantly worse than that on the other instances. This result can be explained by the presence of many plateaus that strongly impede local search.

The performance on instances with edge density equal to 0.036 is also statistically different than that on lower densities and this cannot be explained by the results of position types analysis. The analysis of basins of attraction of local and global minima can shed some light on this point, as it enables us to estimate the probability of reaching a solution to the problem. Basins have been estimated by uniformly sampling the search space with 10^6 samples and by applying a deterministic steepest descent local search. The first outcome of this analysis

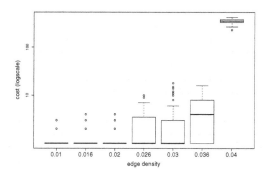

Fig. 4. Box-plots of the performance of randomized non-ascent local search

Table 2. Summary of statistics of relevant characteristics of attractors and their basins

	Edge Density						
	0.010	0.016	0.020	0.026	0.030	0.036	0.040
rGBOA	1.0	0.9561	0.9084	0.5782	0.4831	0.2529	$< 10^{-5}$
Number of cost levels	1	3	4	7	11	14	41
Cost with highest frequency	0	0	0	0	0	2	100
Max cost	0	4	6	12	12	14	100
Median cost	0	0	0	0	2	2	100

is that almost every initial state leads to a different attractor, i.e., almost all basins have size 1.[3] This result can be explained by observing that the problem model induces a search space with many symmetric states, as colors can be permuted [16]. The most relevant statistical characteristics of those attractors are summarized in Table 2. The table reports the fraction of states from which a solution can be reached (*rGBOA*), the number of different levels of cost of the attractors, the most frequent, max and median cost. We can observe that the fraction of attractors corresponding to a solution decreases while edge density increases. The *rGBOA* for instance at density = 0.030 is about 50% of the search space and it halves at density 0.036, till vanishing for the most constrained instance. This difference provides an explanation for the degrading performance of the local search used, that heavily relies on cost decreasing moves. Furthermore, this analysis brings also evidence for the positive correlation between edge density and search space ruggedness.

4.3 Multi-solver Analysis: Tabu Search Configuration Through *F*-Race

As an example of using the multi-solver analysis classes we provide a description of the tuning of tabu search parameters by means of the *F*-Race analysis class.

---

[3] In this analysis the states belonging to the trajectory from the initial state to the attractor are not counted.

Table 3. F-Race results for the configuration of the tabu-list length. Solver values are running-times to a feasible coloring.

Rep-	Solvers						Statistic	p
licate	TS_{5-10}	TS_{10-15}	TS_{15-20}	TS_{20-25}	TS_{25-30}	TS_{30-35}	value	value
01	2.21	21.04	7.62	4.9	4.7	9.68	—	—
02	5.52	3.49	4.3	3.3	8.55	4.64	—	—
03	41.47	12.34	2.9	6.97	12.09	4.43	—	—
04	6.94	2.48	2.29	1.72	4.59	9.74	—	—
05	3.04	2.44	6.13	3.45	7.02	8.71	—	—
06	3.89	7.09	3.02	3.41	2.56	3.29	3.80952	0.577153
07	3.95	4.09	3.1	3.01	2.59	8.99	5.28571	0.382015
			...					
27	28.65	6.73	26.77	7.01	5.9	31.40	11.381	0.044328*
28	7.12	4.3	4.71	3.77	3.5	—	5	0.287297
			...					
37	15.32	25.88	2.99	6.62	8.48	—	9.70811	0.045642*
			...					
40	—	8.35	3.2	15.16	5.3	—	3.99	0.262546

Solvers survived after 40 replicates: TS_{10-15}, TS_{15-20}, TS_{20-25}, TS_{25-30}

Our tabu search implementation employs a dynamic short-term tabu list (called Robust Tabu Search in [9]), so that a move is kept in the tabu list for a random number of iterations in the range $[k_{min}..k_{max}]$. In this example we want to find out the best values of these two parameters among the following set of options: $(k_{min}, k_{max}) \in \{(5,10), (10,15), (15,20), (20,25), (25,30), (30,35)\}$.

As a set of benchmark instances we generate a set of 40 3-colorable equi-partitioned graphs with independent random edge assignment; the graphs have 200 nodes and edge density 0.04. The performance measure employed in this study is the running time needed to reach a feasible coloring.

The results of the F-Race are reported in Table 3 and are those obtained as the output of the report() method of the class Race. We limit the maximum number of replicates to 40 and the confidence level for the Friedman test is 0.95; the first test is performed after 5 replicates. The table summarizes the whole Race procedure, by providing the raw running time values, the value of the statistic employed in the test (the F statistic in the present case) and the p value of the hypothesis testing. For the replicates that lead to discarding one of the candidates, the p value is marked with an asterisk, indicating that the test was significant at the confidence level 0.95. The last line reports the final outcome of the Race and shows the number of replicates performed and the list of solvers that survived the Race.

The results confirm the robustness of employing a dynamic tabu-list. Indeed, only the two most extreme configurations were discarded by the analysis, namely TS_{5-10} and TS_{30-35}.

Of course, these results prompted for additional analysis (for example on different graph sizes), but as in the previous case, this is out of the scope of this presentation since our aim was just to exemplify how to perform an analysis and report its results with a limited effort (see, e.g., [23]).

5   Conclusions

We have presented EASYANALYZER, a software tool for the principled experimental analysis of SLS algorithms. The tool is very general and can be used across a variety of problems with a very limited human effort. In its final version, it will be able to interface natively with a number of development environment, whereas in its current form it is interfaced with EASYLOCAL++, but also with any solver at the price of configuring a command-line interface.

The design of EASYANALYZER deliberately separates the problem-/implementation-specific aspects from the analysis procedures. This allows, for example, to (re)use directly new analyses classes —developed at the framework level— by applying them to all the solvers for which a Solver interface already exists.

We believe that our attempt to define such an environment can be regarded as an initial step toward engineering the experimental analysis of SLS algorithms.

For the future, we will implement the interface modules for the most common environment. We also plan to test EASYANALYZER on more complex problems, with the aim of obtaining also significant results for the research on SLS-based solvers. Finally, we plan to implement other analyses, such as those proposed in [24,19].

References

1. Michel, L., Van Hentenryck, P.: Localizer. Constraints 5(1–2), 43–84 (2000)
2. Van Hentenryck, P., Michel, L. (eds.): Constraint-Based Local Search. MIT Press, Cambridge (MA), USA (2005)
3. Van Hentenryck, P., Michel, L.: Control abstractions for local search. Constraints 10(2), 137–157 (2005)
4. Fink, A., Voß, S.: HotFrame: A heuristic optimization framework. In [25] pp. 81–154
5. Cahon, S., Melab, N., Talbi, E.G.: ParadisEO: A framework for the reusable design of parallel and distributed metaheuristics. Journal of Heuristics 10(3), 357–380 (2004)
6. Voudouris, C., Dorne, R., Lesaint, D., Liret, A.: iOpt: A software toolkit for heuristic search methods. In: Walsh, T. (ed.) CP 2001. LNCS, vol. 2239, pp. 716–719. Springer, Heidelberg (2001)
7. Di Gaspero, L., Schaerf, A.: Writing local search algorithms using EasyLocal++. In[25]
8. Di Gaspero, L., Schaerf, A.: EasyLocal++: An object-oriented framework for flexible design of local search algorithms. Software—Practice and Experience 33(8), 733–765 (2003)
9. Hoos, H., Stützle, T.: Stochastic Local Search Foundations and Applications. Morgan Kaufmann, San Francisco (CA), USA (2005)
10. Halim, S., Yap, R., Lau, H.: Viz: a visual analysis suite for explaining local search behavior. In: Proceedings of the 19th annual ACM symposium on User interface software and technology (UIST '06), pp. 57–66. ACM Press, New York (2006)
11. Lau, H., Wan, W., Lim, M., Halim, S.: A development framework for rapid meta-heuristics hybridization. In: Proceedings of the 28th Annual International Computer Software and Applications Conference (COMPSAC 2004), pp. 362–367 (2004)

12. Gamma, E., Helm, R., Johnson, R., Vlissides, J.: Design Patterns: elements of reusable object-oriented software. Addison-Wesley Publishing, Reading (MA), USA (1995)
13. Fonlupt, C., Robilliard, D., Preux, P., Talbi, E.G.: Fitness landscapes and performance of metaheuristic. In: Voß, S., Martello, S., Osman, I., Roucairol, C. (eds.) Metaheuristics – Advances and Trends in Local Search Paradigms for Optimization, pp. 255–266. Kluwer Academic Publishers, Dordrecht (1999)
14. Pilone, D., Pitman, N.: UML 2.0 in a Nutshell. O'Reilly Media, Inc. Sebastopol (CA), USA (2005)
15. Birattari, M., Stützle, T., Paquete, L., Varrentrapp, K.: A racing algorithm for configuring metaheuristics. In: Proceedings of the Genetic and Evolutionary Computation Conference (GECCO 2002), New York, USA (9-13 July 2002), pp. 11–18. Morgan Kaufmann Publishers, San Francisco (2002)
16. Prestwich, S., Roli, A.: Symmetry breaking and local search spaces. In: Barták, R., Milano, M. (eds.) CPAIOR 2005. LNCS, vol. 3524, Springer, Heidelberg (2005)
17. Jones, T., Rawlins, G.: Reverse hillclimbing, genetic algorithms and the busy beaver problem. In: Genetic Algorithms: Proceedings of the Fifth International Conference (ICGA 1993), San Mateo (CA), USA, pp. 70–75. Morgan Kaufmann Publishers, San Francisco (1993)
18. Birattari, M.: The race package for R. racing methods for the selection of the best. Technical Report TR/IRIDIA/2003-37, IRIDIA, Université Libre de Bruxelles, Brussels, Belgium (2003)
19. Taillard, E.: Few guidelines for analyzing methods. In: Proceedings of the 6th Metaheuristics International Conference (MIC'05), Vienna, Austria (August 2005)
20. Garey, M.R., Johnson, D.: Computers and Intractability: A Guide to the Theory of NP-Completeness. W. H. Freeman, New York (1979)
21. Hertz, A., de Werra, D.: Using tabu search techniques for graph coloring. Computing 39(4), 345–351 (1987)
22. Culberson, J.: Graph coloring page. URL (2004) Viewed: March 2007, Updated: March 2004, http://www.cs.ualberta.ca/~joe/Coloring/
23. Di Gaspero, L., Chiarandini, M., Schaerf, A.: A study on the short-term prohibition mechanisms in tabu search. In: Proc. of the 17th European Conf. on Artificial Intelligence (ECAI-2006) Riva del Garda, Italy pp. 83–87 (2006)
24. Chiarandini, M., Basso, D., Stützle, T.: Statistical methods for the comparison of stochastic optimizers. In: Proceedings of the 6th Metaheuristics International Conference (MIC'05), Vienna, Austria, pp. 189–195 (2005)
25. Voß, S., Woodruff, D. (eds.): Optimization Software Class Libraries. OR/CS. Kluwer Academic Publishers, Dordrecht, the Netherlands (2002)

Mixed Models for the Analysis of Local Search Components

Jørgen Bang-Jensen[1], Marco Chiarandini[1],
Yuri Goegebeur[2], and Bent Jørgensen[2]

[1] Department of Mathematics and Computer Science,
University of Southern Denmark, Odense, Denmark
[2] Department of Statistics,
University of Southern Denmark, Odense, Denmark
{jbj,marco}@imada.sdu.dk, {yuri.goegebeur,bentj}@stat.sdu.dk

Abstract. We consider a possible scenario of experimental analysis on heuristics for optimization: identifying the contribution of local search components when algorithms are evaluated on the basis of solution quality attained.

We discuss the experimental designs with special focus on the role of the test instances in the statistical analysis. Contrary to previous practice of modeling instances as a blocking factor, we treat them as a random factor. Together with algorithms, or their components, which are fixed factors, this leads naturally to a *mixed ANOVA model*. We motivate our choice and illustrate the application of the mixed model on a study of local search for the 2-edge-connectivity problem.

1 Introduction

In recent years, considerable attention has been devoted to the methods for the experimental analysis of heuristic algorithms for optimization. In the case of stochastic local search algorithms [1], experimental analyses have been primarily used to configure and tune the several parameters that are inherent to these algorithms and to show that they are effective. The field of statistics, through its mathematical foundations and well developed tools, has provided the basis for addressing these issues by guaranteeing the reliability and the replicability of the results [2,3,4,5]. This has increased the standard requirements for the acceptance of research papers in the scientific venues of the field. For a typical paper that introduces a new heuristic method for a certain problem and wants to show that it solves it effectively, we might recognize three distinct phases of experimental work.

In a first phase (*Phase I*) a small group of instances with known optima or benchmark solutions are used to evaluate the implementation of new algorithmic ideas, their enhancement and development. This procedure is iterative and interleaves observation and algorithm development. It is continued until some new relevant results arise or some hypothesis, that merits deeper investigation, is identified. This phase is often referred to as *preliminary experimentation* and

T. Stützle, M. Birattari, and H.H. Hoos (Eds.): SLS 2007, LNCS 4638, pp. 91–105, 2007.

there is generally no need to report about it. It is part of the discovery process in the algorithm design.

The second phase (*Phase II*) is designed to assess and validate the results generated in Phase I on a much larger group of test instances with the final aim of inferring the results on a larger scale. In particular, instances must be carefully selected and classified in order to make it possible to distinguish also the differences they might induce on the solvers. Phase II studies might be split into different experiments aimed at emphasizing different aspects. This phase requires a careful planning and the use of methods to identify, emphasize and summarize relevant results. It is important to report about this phase in detail.

The final phase (*Phase III*) consists in running the best heuristic configuration on benchmark instances and in comparing the results with the current state-of-the-art solvers. It is appropriate to report the results of this phase in numerical tables in order to assess the validity and importance of the results attained and to put the work in the context of previous research. If the problem is new, this phase may be substituted with the generation of some new test instances that are to be made publicly available together with numerical results. The information reported must be sufficiently detailed both in terms of computational resources used and solution quality.

Often, in the literature, the distinction between Phase II and Phase III is not well defined while we deem important to maintain these two phases distinct. In Phase III, indeed, the instances available are often not well distributed in the space of all possible instances for the problem. Nevertheless, instance parameters might influence the performance of the algorithms and their study might unveil deeper understanding of the problem. Hence, we advocate, in Phase II, a careful design of the experiments through a factorial[1] organization in which both the algorithm parameters and instance parameters are controlled.

In this paper, we focus on a possible scenario of analysis within Phase II, in which the performance of an algorithm is evaluated on the basis of the quality of the solution it returns. Rather than simply comparing algorithms to determine a single best, like needed in practical applications, we consider studying the importance of algorithmic components, which is an appropriate goal in scientific contributions [6]. In this context, the analysis of variance (ANOVA) offers the mathematical framework for the separation of effects from a large number of results.

Our intention is to examine in detail the application of the ANOVA in this scenario. In particular, we concentrate on the role of the test instances that are randomly sampled from a population. In the literature, instances have been treated as a *blocking factor* [4]. This entails that the results of the analysis are valid only for those specific instances and they should not be claimed valid overall. Although correct for a Phase III, this model is not the best in Phase II, where the intention is to generalize the results over the whole population. In this

---

[1] Throughout the paper we use without difference of meaning the terms factor, commonly used in statistics, and parameter (or component), commonly used in the field of heuristics.

case, instances are more appropriately treated as a *random factor* thus yielding a *mixed model* of fixed and random factors [7,8]. We give analytical evidence of this claim. This model was advocated also in [5] but, to our knowledge, it was not used later.

We illustrate the application of the mixed model analysis in a study of local search components on the graph problem E1-2AUG. This problem consists in augmenting a spanning connected subgraph to a 2-edge-connected graph using only edges from a prescribed set. In commenting the results, we emphasize the differences that arise between the mixed and the blocking model.

We start from basic designs and proceed to the more general case. We restrict ourselves to full factorial designs without considering extensions to fractional factorial designs, which might become necessary with long run-time analyses. In addition, we focus only on parametric methods which are based on few assumptions like the normality and the homoscedasticity of the data. Although these assumptions might not be correct in experiments with heuristics, we advocate their use for the following reasons. The ANOVA by means of the F-test is a well developed and studied technique which is valid with any number of factors. The alternative are rank based tests (such as the Friedman test), but they are only valid for at most two factors, and permutation tests, which are less developed and still under trial for non trivial designs and therefore rarely available in commercially/publicly available statistical software. In addition, the F-test is known to be robust against deviations from the assumptions [8]. Finally, an increase in the number of observations per treatment group helps to become robust against non-normality while data transformations (e.g., log transformation) improves the situation with respect to heteroscedasticity.

Throughout the paper we assume a minimization problem and as measure of solution quality the *percent error* of the approximation of the optimal solution, i.e., $(z-z^*)/z^* \cdot 100$, where z is the observed solution cost in a run of the heuristic on an instance and z^* is the optimal solution cost of the instance.[2] This measure is feasible in our example problem because the optimal solution is known for each instance.

The paper is organized as follows. In Section 2 we formalize the problem of inference and the experimental designs and we provide analytical support for the use of mixed models. In Section 3 we discuss the application example on the E1-2AUG case. Both sections are subdivided into two cases, one dealing with a basic design and another with a more advanced design. We conclude in Section 4 with indications for further extensions of this work.

2   Experimental Design and Statistical Analysis

In the most basic design, the researcher wishes to assess the performance of an heuristic algorithm on a single problem instance π. Since heuristics are, in the

---

[2] This measure does not meet the criteria of invariance for equivalent instances defined by Zemel [9] but we are not concerned with this issue here.

most general case, randomized, their performance Y on one instance is a random variable that might be described by a probability density function $p(y|\pi)$.

More commonly, we aim at drawing conclusions about a certain *class* or *population* of instances Π. In this case, the performance Y of the heuristic on the class Π is described by the probability density function [3,10]

$$p(y) = \sum_{\pi \in \Pi} p(y|\pi)p(\pi), \tag{1}$$

with $p(\pi)$ being the probability of sampling instance π. In other terms, we seek the distribution of Y marginalized over the population of instances.

In experiments, we sample the population of instances and on each sampled instance we collect sample data on the performance of the heuristic algorithm. If on an instance π we run the algorithm r times then we have r replicates of the performance measure Y, denoted $Y_1, \ldots, Y_r$, which are, conditional on the sampled instance, and, given the random nature of the heuristic algorithm, independent and identically distributed (i.i.d.), i.e.

$$p(y_1, \ldots, y_r|\pi) = \prod_{j=1}^{r} p(y_j|\pi). \tag{2}$$

Marginally (over all the instances) the observed performance measures may show dependence, as is easily seen from

$$p(y_1, \ldots, y_r) = \sum_{\pi \in \Pi} p(y_1, \ldots, y_r|\pi)p(\pi). \tag{3}$$

This model can be easily extended to the case where several algorithms are applied to the same instance by incorporating fixed effects in the conditional structure of (2). We now illustrate with two cases how the model (3) entails naturally a mixed model for the description of the effects in data stemming from experiments in which optimization heuristics are compared.

2.1   Case 1: Two Factors Mixed Design

Our first experiment consists of h heuristic *algorithms* evaluated on p *instances* randomly sampled from a class Π. The experiment is performed as follows. In a first stage an instance is sampled from a population of instances. Next, each algorithm is run r times on the instance. Given the stochastic nature of the algorithms, this produces, *conditional on the instance,* r i.i.d. replications of the performance measure. We use Y_{ijk} to denote the random performance measure obtained in replication k of algorithm j on instance i.

The algorithms included in the study are the ones in which we are particularly interested, and hence they can be considered as levels of a fixed factor. The instances on the other hand are drawn randomly from some population of instances, and the interest is in inferring about this global population of instances, not just those included in the study. Hence, we assume that the performance measure can be decomposed into the following *mixed effects ANOVA model*

$$Y_{ijk} = \mu + \alpha_j + \tau_i + \gamma_{ij} + \varepsilon_{ijk}, \tag{4}$$

where

μ is an overall performance level common to all observations,
α_j is a fixed effect due to the algorithm j,
τ_i is a random effect associated with instance i,
γ_{ij} is a random interaction between instance i and algorithm j,
ε_{ijk} is a random error for replication k of algorithm j on instance i.

The assumptions imposed on the random elements are

τ_i are i.i.d. $N(0, \sigma_\tau^2)$,
γ_{ij} are i.i.d. $N(0, \sigma_\gamma^2)$,
ε_{ijk} are i.i.d. $N(0, \sigma^2)$,
the τ_i, γ_{ij} and ε_{ijk} are mutually independent random variables.

Note that the postulated mixed effects ANOVA model satisfies the structure of the conditional and marginal models given by (2) and (3). In particular, the conditional distribution of the performance measure given the instance and the instance-algorithm interaction is given by

$$Y_{ijk} | \tau_i, \gamma_{ij} \sim N(\mu + \alpha_j + \tau_i + \gamma_{ij}, \sigma^2), \quad i = 1, \ldots, p, \ j = 1, \ldots, h, \ k = 1, \ldots, r.$$

Furthermore, conditional on the random effects τ_i and γ_{ij}, $i = 1, \ldots, p$, $j = 1, \ldots, h$, all responses are independent. Integrating out the random effects we obtain the unconditional model

$$Y_{ijk} \sim N(\mu + \alpha_j, \sigma^2 + \sigma_\tau^2 + \sigma_\gamma^2), \quad i = 1, \ldots, p, \ j = 1, \ldots, h, \ k = 1, \ldots, r.$$

Moreover, the use of random instance effects and random instance-algorithm interactions yields dependency between the performance measurements obtained at a specific instance. In particular we have

$$\text{Cov}(Y_{ijk}, Y_{i'j'k'}) = \begin{cases} \sigma^2 + \sigma_\tau^2 + \sigma_\gamma^2, & \text{if } i = i', \ j = j', \ k = k', \\ \sigma_\tau^2 + \sigma_\gamma^2, & \text{if } i = i', \ j = j', \ k \neq k', \\ \sigma_\tau^2, & \text{if } i = i', \ j \neq j', \\ 0, & \text{otherwise.} \end{cases} \tag{5}$$

The parameters σ^2, σ_τ^2 and σ_γ^2 determine the variance of the individual Y_{ijk} as well as the covariance between the Y_{ijk}, and therefore are called the *variance components*.

In the literature the factor *instance* has been treated as a blocking factor. This entails that the analysis remains valid only in the specific instances and should not be generalized to the whole class of possible instances. Further, when the instances are considered as a blocking factor in model (4), the instance effects and instance-algorithm interactions are fixed parameters, yielding

$$Y_{ijk} \sim N(\mu + \alpha_j + \tau_i + \gamma_{ij}, \sigma^2), \quad i = 1, \ldots, p, \ j = 1, \ldots, h, \ k = 1, \ldots, r.$$

with Y_{ijk} being independent, i.e., unlike (4), the model does not take dependencies arising from applying algorithms to the same instances into account.

The mixed model (4) with its assumptions forms the natural basis for testing hypotheses about both fixed and random factors, and their interactions. Concerning the fixed factors, the interest is usually in testing whether there is a difference between the factor level means $\mu + \alpha_1, \ldots, \mu + \alpha_h$. Formally, one tests the hypothesis

$$H_0 : \alpha_1 = \alpha_2 = \ldots = \alpha_h = 0,$$
$$H_1 : \text{ at least one } \alpha_j \text{ not equal to } 0.$$

For the random effects, tests about the particular levels included in the study are meaningless. Instead we test hypotheses about the variance components σ_τ^2 and σ_γ^2, reflecting that ultimate interest is in the whole population of instances:

$$H_0 : \sigma_\tau^2 = 0, \qquad \text{and} \qquad H_0 : \sigma_\gamma^2 = 0,$$
$$H_1 : \sigma_\tau^2 > 0, \qquad \qquad \qquad H_1 : \sigma_\gamma^2 > 0,$$

respectively. We refer to [7,8,11] for a discussion on how the analysis of variance is conducted in the case of mixed models.

2.2 Case 2: General Mixed Design

In this case the researcher wishes to assess how the performance measure Y is affected by several parameters of the heuristics and of the instances. Ideally, we fix those parameters that are most important and that we can control, and randomize those properties that we do not understand or cannot control. The parameters controlled may be both categorical or numerical. We consider the following setting:

- Factors $A_1, \ldots, A_a$ represent the parameters of the heuristics. Each combination of these factors gives rise to an instantiated algorithm.
- Factors $B_1, \ldots, B_b$ represent the parameters of the instances or the *stratification factors* of the whole space of instances. Each combination of the factors $B_1, \ldots, B_b$ gives rise to a different class of instances Π_m.
- From each class of instances Π_m, p instances are sampled randomly and on each of them each instantiated algorithm is run once. We could also run each algorithm r times and extension of the analysis to this case would be straightforward.

The factors A_i, $i = 1, \ldots, a$, and B_j, $j = 1, \ldots, b$, are fixed factors and the factor instance is a random factor. Since the instances within each class Π_m are different the design is *nested*. This yields a linear mixed model that can be written as

$$Y_{ABk} = \mu + \alpha_A + \beta_B + \tau_{B(k)} + \varepsilon_{ABk}$$

with

$A = \{i_1, \ldots, i_a\}$, an index set referring to the levels of the algorithmic factors,

$B = \{j_1, \ldots, j_b\}$, an index set referring to the levels of the instance factors,

α_A: a vector containing the factor levels of the algorithmic factors indexed by A,

β_B: a vector containing the factor levels of the instance factors indexed by B,

$\tau_{B(k)}$: the random effect of instance k in setting B of the instance factors,

ε_{ABk}: a random error term.

3   Augmenting a Tree to a 2-Edge-Connected Graph

The application example is extracted from a study on heuristic and exact algorithms for the so-called E1-2AUG problem. We refer to [12] for the full details. Here, we focus on an intermediate result and the relative Phase II experiments. We do not consider therefore state-of-the-art algorithms. We first describe shortly the problem using graph theory notation (see [13] for an introduction). Then we describe briefly the local search algorithms and the instance classes studied in [12].

3.1   Definitions and Problem Formulation

An edge uv in a connected graph $G = (V, E)$ is a *bridge* if we can partition V into two sets $S, V - S$ so that uv is the only edge from E with endpoints in both S and $V - S$. A graph is *2-edge-connected* if it is connected and has no bridges.

In an instance of the E1-2AUG problem, we are given an undirected 2-edge-connected graph $G = (V, E)$, a fixed spanning connected subgraph of G, $S = (V, F)$, and a non-negative weight function ω on $E' = E \setminus F$. The task is finding a subset X of E' of minimal total weight so that $A(G) = (V, F \cup X)$ is 2-edge-connected.

We consider only the case in which the graph G is a simple graph and S is a tree. An edge $uv \in E$ which is not in S is said to *cover* those edges of S which correspond to the unique uv-path P_{uv} in S. We assume that every edge uv in F is covered by at least two edges in E'. We call a subset X of E' a *proper augmentation* of S if $A(G) = (V, F \cup X)$ is 2-edge-connected.

Every optimal augmentation X is minimal, that is, no edge can be deleted from X without leaving at least one edge of S uncovered. If a given augmentation is not minimal it can be made so by means of a *trimming* procedure that removes edges from X without leaving any edge of S uncovered.

It can be noted that the E1-2AUG problem is a special case of the general set covering problem [14]. Let us associate with each edge $uv \in E'$ the subset W_{uv} of edges in S corresponding to the unique uv-path P_{uv} in S and give this set the same weight as the edge uv. Then, $X \subseteq E'$ is a proper augmentation if and only if the corresponding collection of subsets $\{W_e : e \in X\}$ covers every element in the set F.

A heuristic approach to the E1-2AUG problem has been proposed by Raidl and Ljubic [15] who designed an evolutionary local search algorithm. Their local search consists in picking an edge in X randomly and checking whether it can be removed. We studied more aggressive local search schemes.

3.2   Local Search Schemes

We assume that the local search is a first improvement procedure that starts from a minimal proper augmentation.

Addition neighborhood. Neighboring augmentations are obtained by adding l edges from $E' - X$ and trimming the resulting augmentation. We restrict the examination of the neighborhood to only a choice of edges to add (called candidate superior edges). This neighborhood is inspired by the algorithm for set covering by Marchiori and Steenbeck [1, pag. 494].

Destruct-reconstruct neighborhood. Neighboring augmentations are obtained by removing l edges from the current augmentation and reconstructing the resulting improper augmentation by means of the greedy set covering heuristic by Chvatal [16, pag. 1035].

Shortest path neighborhood. This neighborhood exploits the graph formulation. It uses the following observation: if X is a proper augmentation which is minimal and $Y = X - uv$ for some edge $uv \in X$, then, using a shortest path calculation in a suitable digraph, we can find a minimum cost set of new edges $Z \subseteq E' - X - uv$ so that $(X - uv) \cup Z$ is again a proper (not necessarily minimum) augmentation [12]. More definitely, for $l = 1$ a step in the neighborhood consists of deleting an edge uv, finding Z via a shortest path calculation and then trimming $(X - uv) + Z$. For $l > 1$ the process is more complicated and we refer to [12] for its treatment.

3.3   Problem Instances

When instances are generated randomly, particular attention must be given to possible biases. As an example, the possible non-isomorphic graphs of size 800 are more than those of size 200 and hence they should be given more probability to appear. In practice, one proceeds by stratifying the instance space, that is, partitioning the instance space by means of some instance parameters. The analysis is then restricted to the specific classes created. In the case of the E1-2AUG problem we consider three stratification parameters: *type*, *size* and *edge density*. The type distinguishes between *uniform* random graphs (Type U) and *geometric* random graphs (Type G). The former are generated by including each of the $\binom{n}{2}$ possible edges independently with probability p and assigning on these edges integer weights randomly chosen from the interval $(1, 10000)$. The latter are generated from points in a two dimensional grid with random, integer coordinates in $[1, 10000]$. Edges with weights equal to the Euclidean distance between their incident points, rounded to the closest integer, are included if this distance is less than an integer parameter d. In both types of graphs, the spanning tree S is chosen randomly among those of minimal weight.

3.4 Experimental Analysis

We study the two design cases introduced in Section 2. We further subdivide case 1 in replicated and unreplicated design.

Case 1: Two factors mixed design

Replicated. We aim at comparing the performance of the shortest path neighborhood at different values of l, over an instance class, namely instances of type G, size 400 and edge density 0.5. Hence we consider the following factors:

- algorithm: three algorithms, namely the l-shortest path neighborhood with $l = \{1, 3, 5\}$ starting from the same solution;
- instance: 10 instances randomly sampled from the class defined.

We collect 3 runs per algorithm on each instance, thus yielding in total 90 observations. A way to inspect the data is to plot the percentage error of the algorithms within each instance as in Figure 1. As is clear from this figure, the instances cause shifts in the performance and also affect the effect of switching algorithms. We hence expect an important instance effect as well as important algorithm-instance interaction.

The statistical analysis of this model can be performed with the SAS procedure Mixed. All results reported in this paper were obtained by fitting the model with the restricted maximum likelihood (REML) method. Relevant questions for this design are

- Is there an instance effect, i.e., do the instances contribute significantly to the variability of the responses?
- Do the mean performances of the algorithms differ? If yes, how different are they?
- Do the instance-algorithm interactions contribute significantly to the variability of the responses?

The results of the analysis are displayed in Table 1. The section 'Covariance Parameter Estimates' reports the estimated variances for the instance and the instance-algorithm interaction random effects which are $\hat{\sigma}_\tau^2 = 3.27$ and $\hat{\sigma}_\gamma^2 = 0.61$, respectively. The last columns give an indication of the acceptance or rejection of the null hypothesis that the respective variance component is zero. The column Pr Z gives the p-value of a one-sided normal distribution based test, i.e., $P(Z > z^*)$, where $Z \sim N(0, 1)$ and z^* the observed value of the test statistic (here a standardized variance component, column Z Value). We assume a significance level of 5%. As is clear, both the instances and the instance-algorithm interactions contribute significantly to the variability of the performance measure, and hence, given (5), measurements obtained on a particular instance show

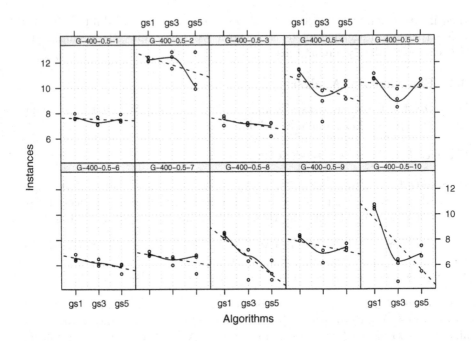

Fig. 1. The data in the replicated design. The three algorithms, gs1, gs3 and gs5, correspond to greedy covering construction (g) followed by shortest path local search (s) with $l = \{1, 3, 5\}$. A local regression line (dashed) and a least square linear regression line are superimposed.

dependence. The section 'Type 3 Tests of Fixed Effects' in Table 1 analyzes the fixed effects. The observed F statistic is 8.26 (column F value) on 2 and 18 degrees of freedom (DF) with p-value of 0.0029 (Pr > F). We can conclude that at least one of the algorithms considered differs significantly from the others. The section 'Least Squares Means' reports the generalized least squares estimates of the fixed effects model parameters from which we obtain the point estimates for the mean performance of the algorithms, i.e., $\mu + \alpha_1$, $\mu + \alpha_2$ and $\mu + \alpha_3$. The last column in this section gives the p-value for the rather uninformative hypotheses that the mean performances are zero. Pairwise tests of these means (not displayed), corrected for simultaneous testing, indicate that algorithm 1 differs significantly from algorithms 2 and 3 in mean performance, while algorithms 2 and 3 do not show significant differences. We conclude that the shortest path local search attains better performance with $l > 1$.

It is instructive to compare the results obtained here under a random effects model with those obtained by considering instances as blocks. In the latter case, the test for algorithmic differences would have been performed relative to the mean square error, and not relative to the instance-algorithm interaction mean square. In case of blocking, the F test has 60 denominator degrees of freedom, compared to 18 under a mixed model, and hence, for the same significance level it will reject sooner.

Table 1. Replicated case: mixed model analysis of the main effects

```
                    The Mixed Procedure
              Covariance Parameter Estimates
                          Standard           Z
Cov Parm        Estimate      Error      Value         Pr Z
inst              3.2662     1.6593       1.97       0.0245
algo*inst         0.6112     0.2505       2.44       0.0073
Residual          0.4095    0.07476       5.48       <.0001
              Type 3 Tests of Fixed Effects
              Num      Den
Effect         DF       DF    F Value     Pr > F
algo            2       18       8.26     0.0029
                  Least Squares Means
                             Standard
Effect   algo       Estimate     Error     DF    t Value    Pr > |t|
algo     gs1          9.0216    0.6336   11.6      14.24      <.0001
algo     gs3          7.5883    0.6336   11.6      11.98      <.0001
algo     gs5          7.7467    0.6336   11.6      12.23      <.0001
```

Unreplicated. In this case, we aim at comparing again the three algorithms above with a set of 90 total experiments, as it might be imposed, for example, by computational time considerations. Yet, we decide now to use 30 instances sampled from the class and perform only one single run per algorithm on each of them.

Clearly, the mixed model can be applied also in the unreplicated case. However, it can be proven that when $r = 1$, the error variance σ^2 and interaction variance σ_γ^2 of model (4) cannot be separately identified. We therefore consider a restricted version of (4) obtained by omitting the random interaction terms. The estimation results are summarized in Table 2. Again we find that the variability due to the sampling of instances contributes significantly to the variance of the performance measure and that at least two of the algorithms considered differ in their mean performance. Note that in case of the unreplicated design the standard error of the mean algorithmic performance is smaller than in the replicated case (0.3670 in the unreplicated case versus 0.6336 in the replicated case), an observation which is consistent with [10]. Moreover, when a bound on the total number of experiments is present, the unreplicated case yields a more powerful test for differences between the levels of the fixed factor. The results for the random factor are ambiguous in this respect, and depend on the magnitudes of the variance components. However, as mentioned above, the unreplicated case does not allow estimation of the algorithm-instance interaction variance.

Case 2: General mixed design

We aim at a detailed understanding of the influence of local search components in the performance with respect to different instance features. We consider the following factors

Table 2. Unreplicated case: mixed model analysis of the main effects

The Mixed Procedure
Covariance Parameter Estimates

Cov Parm	Estimate	Standard Error	Z Value	Pr Z
inst	3.2138	0.9177	3.50	0.0002
Residual	0.8258	0.1533	5.39	<.0001

Type 3 Tests of Fixed Effects

Effect	Num DF	Den DF	F Value	Pr > F
algo	2	58	37.92	<.0001

Least Squares Means

Effect	algo	Estimate	Standard Error	DF	t Value	Pr > |t|
algo	gs1	9.0959	0.3670	38.4	24.79	<.0001
algo	gs3	7.1470	0.3670	38.4	19.48	<.0001
algo	gs5	7.5895	0.3670	38.4	20.68	<.0001

- initial: the starting solution generated by three different construction heuristics {greedy_cov, lightest_add, shortest_path} (see [12] for a description);
- neighborhood: the three local search schemes described above, which we call in the order: {l-add, l-cov, l-sp};
- l: the parameter that determines the extension of the neighborhood with values in {1,3,5}.

The first two factors are categorical, the last one is numerical but we treat it as categorical because this removes the constraint that the change in the mean response is the same for shifting l from 1 to 3 as for shifting l from 3 to 5. We are also interested in studying the instance features. The factors describing them are

- type: equal to {U, G};
- size: equal to {200, 400, 800};
- dens: equal to {0.1,0.5,0.9}.

Both the factors concerning the algorithms and those concerning the instances are fixed factors. Each combination of the three instance factors gives rise to a class from which we sample 5 instances. The factor inst is, hence, a random factor. The experiment has $3^5 \cdot 2 \cdot 5 = 2430$ experimental units. Interactions between the fixed effects are detectable up to the 6th level.

Table 3 shows the ANOVA results. We conclude that the variability of the instances within each class contributes significantly to the variance of the performance measure. Moreover, we observe that the instance factors size and dens

Table 3. General unreplicated case: mixed model analysis of the main effects. Least squares means are truncated to only the best algorithmic configurations.

The Mixed Procedure				
Cov Parm	Estimate	Error	Value	Pr Z
inst	3.7623	0.7095	5.30	<.0001
Residual	19.5141	0.5767	33.84	<.0001

Type 3 Tests of Fixed Effects				
	Num	Den		
Effect	DF	DF	F Value	Pr > F
initial	2	2290	323.25	<.0001
neighborhood	2	2290	105.18	<.0001
l	2	2290	42.72	<.0001
type	1	80	105.65	<.0001
size	1	80	1.04	0.3105
dens	2	80	0.37	0.6948
initial*neighborhood	4	2290	50.22	<.0001
initial*l	4	2290	61.86	<.0001
initial*type	2	2290	1248.31	<.0001
size*initial	2	2290	6.86	0.0011
initial*dens	4	2290	2.22	0.0645
neighborhood*l	4	2290	25.20	<.0001
neighborhood*type	2	2290	92.08	<.0001
size*neighborhood	2	2290	1.19	0.3056
neighborhood*dens	4	2290	1.37	0.2407
l*type	2	2290	93.49	<.0001
size*l	2	2290	3.38	0.0341
l*dens	4	2290	0.17	0.9544
size*type	1	80	3.11	0.0817
type*dens	2	80	0.09	0.9113
size*dens	2	80	0.37	0.6939
initial*neighborho*l	8	2290	17.88	<.0001

Least Squares Means					
		Standard			
algo	Estimate	Error	DF	t Value	Pr > |t|
greedy_cov.l-add.1	3.1247	0.5086	1336	6.14	<.0001
greedy_cov.l-add.3	3.2907	0.5086	1336	6.47	<.0001
greedy_cov.l-add.5	3.4624	0.5086	1336	6.81	<.0001
greedy_cov.l-cov.1	6.4922	0.5086	1336	12.77	<.0001
greedy_cov.l-cov.3	6.3530	0.5086	1336	12.49	<.0001
greedy_cov.l-cov.5	6.2631	0.5086	1336	12.32	<.0001
greedy_cov.l-sp.1	7.6148	0.5086	1336	14.97	<.0001
greedy_cov.l-sp.3	5.9073	0.5086	1336	11.62	<.0001
greedy_cov.l-sp.5	6.0453	0.5086	1336	11.89	<.0001
shortest_path.l-add.1	6.3345	0.5086	1336	12.46	<.0001
shortest_path.l-add.3	6.2280	0.5086	1336	12.25	<.0001
shortest_path.l-add.5	5.7602	0.5086	1336	11.33	<.0001
...					

do not have a significant main effect on the quality of the solutions found while there is a main effect of the type of instance. It would be therefore appropriate to split the analysis in the two types of instances. All the local search components show instead significant effects. Looking at the least squares mean estimates we learn that in the shortest path and in the destruct-reconstruct neighborhood a value $l > 1$ yields better performance while for the addition neighborhood we have the opposite result. This explains the presence of a significant interaction between l and neighborhood type. This result holds on both instance types. Deeper investigation on the effects of the instance type unveils that: on instances of type G all algorithms attain in general better percentage errors (main effect) and the order of the configurations changes (interaction effect), although not for the addition neighborhood that remains consistently the best.

Finally, we would like to emphasize, in this case, the difference with a blocked experiment. In a blocked experiment with the instances as blocks, one would consider all combinations of the factor levels in all blocks. But this is clearly not feasible here because we do not have all possible factor level combinations for a particular instance. Rather, instances are *nested* within the different instance factor level combinations.

4   Conclusions and Directions for Further Research

We examined the application of the analysis of variance in the study of components of heuristic algorithms. We showed that a *mixed model* of fixed factors, the algorithm components, and a random factor, the instances, is more appropriate than the blocking model to generalize the results to the class of instances. We discussed the implications of the two different models both analytically and in a practical example. In addition, we emphasized the importance of organizing the experiments also with respect to instance parameters. This methodology is suitable and helpful in scientific studies and we suggest its use in addition to the widespread practice of testing on benchmark instances and reporting the results in numerical tables.

There are a number of extensions that we could not include here for reason of space. Diagnostic plots can be obtained also for the mixed models and we used them to check that no transformation was needed in our data. Pairwise comparisons in a mixed model are possible and provide further detail in the analysis beside the least squares mean estimates. The extension to a bivariate analysis that takes into account both solution-cost and run-time is instead not trivial and it might be the focus of future study.

Acknowledgment. The authors wish to thank Peter Morling for the implementation of the local search algorithms.

References

1. Hoos, H., Stützle, T.: Stochastic Local Search: Foundations and Applications. Morgan Kaufmann, San Francisco, CA, USA (2004)
2. Barr, R., Golden, B., Kelly, J., Resende, M., Stewart, W.: Designing and reporting on computational experiments with heuristic methods. Journal of Heuristics 1(1), 9–32 (1995)
3. McGeoch, C.C.: Toward an experimental method for algorithm simulation. INFORMS Journal on Computing 8(1), 1–15 (1996)
4. Rardin, R.L., Uzsoy, R.: Experimental evaluation of heuristic optimization algorithms: A tutorial. Journal of Heuristics 7(3), 261–304 (2001)
5. Coffin, M., Saltzman, M.J.: Statistical analysis of computational tests of algorithms and heuristics. INFORMS Journal on Computing 12(1), 24–44 (2000)
6. Hooker, J.N.: Testing heuristics: We have it all wrong. Journal of Heuristics 1, 32–42 (1996)
7. Molenberghs, G., Verbeke, G. (eds.): Linear Mixed Models in Practice - A SAS-Oriented Approach. Springer, Heidelberg (1997)
8. Montgomery, D.C.: Design and Analysis of Experiments, 6th edn. John Wiley & Sons, Chichester (2005)
9. Zemel, E.: Measuring the quality of approximate solutions to zero-one programming problems. Mathematics of operations research 6(3), 319–332 (1981)
10. Birattari, M.: On the estimation of the expected performance of a metaheuristic on a class of instances. how many instances, how many runs? Tech. Rep. TR/IRIDIA/2004-01, IRIDIA, Université Libre de Bruxelles, Brussels, Belgium (2004)
11. Molenberghs, G., Verbeke, G.: Models for Discrete Longitudinal Data. Springer, Heidelberg (2005)
12. Bang-Jensen, J., Chiarandini, M., Morling, P.: A computational investigation on heuristics for 2-edge connectivity augmentation. Submitted to journal (2007)
13. Diestel, R.: Graph Theory, 2nd edn. electronic edn. Springer-Verlag, New York, Berlin (2000)
14. Conforti, M., Galluccio, A., Proietti, G.: Edge-connectivity augmentation and network matrices. In: Hromkovič, J., Nagl, M., Westfechtel, B. (eds.) WG 2004. LNCS, vol. 3353, pp. 355–364. Springer, Heidelberg (2004)
15. Raidl, G.R., Ljubic, I.: Evolutionary local search for the edge-biconnectivity augmentation problem. Information Processing Letters 82(1), 39–45 (2002)
16. Cormen, T., Leiserson, C., Rivest, R.: Introduction to algorithms, 2nd edn. MIT Press, Cambridge (2001)

An Algorithm Portfolio for the Sub-graph Isomorphism Problem

Roberto Battiti and Franco Mascia

Università degli Studi di Trento, Trento, Italy
{battiti,mascia}@dit.unitn.it

Abstract. This work presents an algorithm for the sub-graph isomorphism problem based on a new pruning technique for directed graphs. During the tree search, the method checks if a new association between two vertices is compatible by considering the structure of their local neighborhoods, represented as the number of limited-length paths of different type originating from each vertex. In addition, randomized versions of the algorithms are studied experimentally by deriving their runtime distributions. Finally, algorithm portfolios consisting of multiple instances of the same randomized algorithm are proposed and analyzed. The experimental results on benchmark graphs demonstrate that the new pruning method is competitive w.r.t. recently proposed techniques. Significantly better results are obtained on sparse graphs. Furthermore, even better results are obtained by the portfolios, when both the average and standard deviation of solution times are considered.

1 Introduction

The sub-graph isomorphism problem, a.k.a. graph pattern matching, consists of determining if an isomorphic image of a graph is present in a second graph. The problem, or relaxed versions thereof, appears in significant applications, ranging from computer vision, structural pattern recognition, chemical documentation, computer-aided design, and visual languages, see for example [1,2,3,4] for references.

Let $G_1(V_1, E_1)$ and $G_2(V_2, E_2)$ be two graphs, V and E being their vertices and edges, respectively. A sub-graph isomorphism is a bijective function $M : V_1 \rightarrow V_2' \subseteq V_2$ having the following property: $(u, v) \in E_1 \Leftrightarrow (M(u), M(v)) \in E_2' \subseteq E_2$, where E_2' contains the edges induced by the vertices in V_2'. Let's note that another definition has been used in some papers, for example [3], where existence of an arc in G_1 implies existence of an arc $(M(u), M(v))$ in G_2, but not *vice versa*, there can be arcs in E_2' which do not correspond to arcs in G_1.

The original motivation for this work is double. First, we investigate whether the adoption of a portfolio approach produces better results by considering more instances of the same algorithm running in a time-sharing fashion. Second, we experiment with novel pruning techniques based on the local structure around the next node to be associated.

T. Stützle, M. Birattari, and H.H. Hoos (Eds.): SLS 2007, LNCS 4638, pp. 106–120, 2007.

In the following sections, the existing state-of-the-art approaches are briefly reviewed in Section 2, then our new pruning technique based on paths compatibility is explained in Section 3. The computational experiments to assess the efficacy and efficiency of the new pruning proposal are presented in Section 4 for the deterministic algorithms, and in Section 5 for the randomized versions. The motivation for using portfolios and the proposal is explained in Section 6, and the corresponding computational results are presented in Section 7.

2 Existing Approaches

The sub-graph isomorphism problem is NP-hard [5], and previous approaches for its solution include [1,2,3,4]. A recent algorithm appropriate for matching large graphs encountered in relevant applications is proposed in [4]. The proposed method VF2 is an exact algorithm for the sub-graph isomorphism problem, which explores the search graph by means of a depth-first-search and which uses new pruning techniques to reduce the size of the generated solution tree. The effectiveness of VF2 is assessed in the cited paper, which contains also experimental comparisons with Ullmann [1] and Nauty [6] algorithms.

Let's introduce the notation used to explain VF2 and our novel proposal. Let $M \subset V_1 \times V_2$ be the isomorphism, and M_s the mapping at state s in the state space representation. A mapping is developed by adding a new pair of nodes (v_1, v_2) at each step, and the state s is given by the current set of associations between nodes of G_1 and nodes of G_2. $M_1(s)$ and $M_2(s)$ are the set of vertices $v_1 \in V_1, v_2 \in V_2$ such that $(v_1, v_2) \in M_s$ and $G_1(s)$ and $G_2(s)$ the sub-graphs induced by these sets. Let $T_1^{in}(s)$ and $T_1^{out}(s)$ be the set of vertices adjacent from and to the vertices in $M_1(s)$, but not yet in the partial mapping $M_1(s)$,

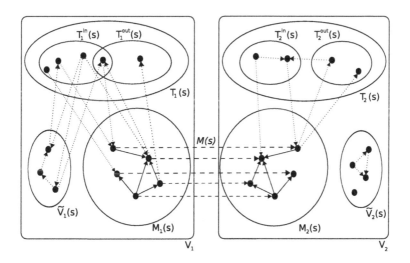

Fig. 1. Partial mapping M_s and sets in VF2

1. MATCH (G_1, G_2, s)
2. ┌ **if** M_s covers all the vertices of G_1 **then**
3. │   **return** M_s
4. │ **else**
5. │   ┌ **foreach** $(v_1, v_2) \in P(s)$ **do**
6. │   │   ┌ **if** COMPATIBLE(v_1, v_2) **then**
7. │   │   │   ┌ $s' \leftarrow s \cup (v_1, v_2)$
8. │   │   └   └ MATCH (G_1, G_2, s')
9. │   └ **done**
10. └ **return** no match found

Fig. 2. A generic back-tracking scheme for the sub-graph isomorphism problem. The function COMPATIBLE determines the pruning and it depends on the specific algorithm.

and $T_1(s) = T_1^{in}(s) \cup T_1^{out}(s)$. The set of the vertices $\tilde{V}_1 = V_1 - M_1(s) - T_1(s)$ is the set of vertices $u \in V_1$ not connected to vertices belonging to the mapping.

Fig. 1 shows the sets described above, highlighting the connections among the induced sub-graphs $G_1(s)$ and $G_2(s)$ with solid arcs, the partial mapping with dashed arcs, and connections with the terminal sets with dotted arcs.

Fig. 2 shows the back-tracking algorithm which implements the depth-first-search. If the partial mapping M_s covers all the vertices of G_1, the goal is reached, otherwise the depth-first-search goes deeper in the search tree and tries to add a new pair to the current state s.

To reduce as much as possible the CPU time, by an appropriate ordering the algorithm never visits the same state twice. The search space reduction w.r.t. the complete search tree is determined by the pruning technique. Pruning acts by controlling that a candidate pair $p = (v_1, v_2)$ selected from the set of candidate pairs $P(s)$ survives the test executed by the COMPATIBLE routine. If COMPATIBLE returns false, the addition of the new pair is doomed to failure and the sub-tree is pruned.

In detail, the COMPATIBLE routine for VF2 works as follows. The candidate pairs $(v_1, v_2) \in P(s)$ are selected with priority to the nodes adjacent *from* the vertices already in the mapping, i.e., $v_1 \in T_1^{out}(s)$ and $v_2 \in T_2^{out}(s)$. If there are no such vertices, the pairs of the ones adjacent *to* the vertices already in the mapping are selected. If a graph has more than one connected component such couples could not exists, and in this case the "less constrained" vertices belonging to $\tilde{V}_1(s)$ and $\tilde{V}_2(s)$ are considered.

Let us now introduce the sets of predecessors and successors of the current node: $\mathsf{Pred}(G, v) = \{u \in V | (u, v) \in E\}$ and $\mathsf{Succ}(G, v) = \{u \in V | (v, u) \in E\}$. The COMPATIBLE routine performs each of the following tests in order, stopping early if at least one test fails. The first test checks if the partial mapping extended with the additional association $(M_s \cup (v_1, v_2)$, where $(v_1, v_2) \in P(s))$ is still a valid isomorphism: for all nodes already in the partial mapping, edges to (from) the last nodes considered for addition must be preserved by the extended mapping: if an edge is present in the graph induced by $M_1(s) \cup v_1$

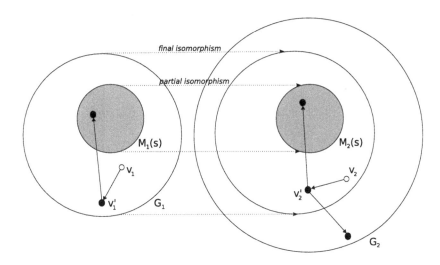

Fig. 3. Example of the additional checks executed by VF2. A node in G_1 will have to be mapped to a compatible node in G_2 in the future steps. If no compatible node in G_2 is available the partial mapping is doomed.

the corresponding edge must be present in the graph induced by $M_2(s) \cup v_2$, and *vice versa*.

$$\forall v_1' \in M_1(s) \quad (v_1', v_1) \in E_1 \Rightarrow (M_s(v_1'), v_2) \in E_2 \tag{1}$$

$$\forall v_1' \in M_1(s) \quad (v_1, v_1') \in E_1 \Rightarrow (v_2, M_s(v_1')) \in E_2 \tag{2}$$

$$\forall v_2' \in M_2(s) \quad (v_2', v_2) \in E_2 \Rightarrow (M_s^{-1}(v_2'), v_1) \in E_1 \tag{3}$$

$$\forall v_2' \in M_2(s) \quad (v_2, v_2') \in E_2 \Rightarrow (v_1, M_s^{-1}(v_2')) \in E_1 \tag{4}$$

If the previous checks give the green light to extend the mapping, the following additional checks are performed to prune the search tree, trying to find incompatibilities between branches of the two graphs that could arise in the future steps. The tests count number of nodes with different connectivity structure w.r.t. M_1 in G_1 and make sure that *at least* the same number of nodes with compatible connectivity structure is available in G_2. Otherwise, for sure the mapping cannot be completed in the future steps. To follow the different cases it may be useful to consider the example in Fig. 3, related to the check in eqn. 5. The node external to M_1 is a successor of v_1 and has at least one incoming arc to M_1. If the mapping is to be completed, at least one node in G_2 external to M_2 with compatible edges has to be present. Again, given a number of nodes with a certain connectivity in $G_1 \setminus M_1$, at least the same number of nodes with compatible connectivity has to be present in $G_2 \setminus M_2$. Let's note that, in addition to the required edges, some additional edges may be present in G_2 because only a subset of its nodes will be covered by the final mapping. The different cases consider all possible directions for the edges (in-in, in-out, out-out, out-in) and finally the case of successors and predecessors without edges to or from the current M_1.

$$|T_1^{in}(s) \cap \mathsf{Succ}(G_1, v_1)| \leq |T_2^{in}(s) \cap \mathsf{Succ}(G_2, v_2)| \tag{5}$$

$$|T_1^{in}(s) \cap \mathsf{Pred}(G_1, v_1)| \leq |T_2^{in}(s) \cap \mathsf{Pred}(G_2, v_2)| \tag{6}$$

$$|T_1^{out}(s) \cap \mathsf{Succ}(G_1, v_1)| \leq |T_2^{out}(s) \cap \mathsf{Succ}(G_2, v_2)| \tag{7}$$

$$|T_1^{out}(s) \cap \mathsf{Pred}(G_1, v_1)| \leq |T_2^{out}(s) \cap \mathsf{Pred}(G_2, v_2)| \tag{8}$$

$$|\tilde{V}_1(s) \cap \mathsf{Succ}(G_1, v_1)| \leq |\tilde{V}_2(s) \cap \mathsf{Succ}(G_2, v_2)| \tag{9}$$

$$|\tilde{V}_1(s) \cap \mathsf{Pred}(G_1, v_1)| \leq |\tilde{V}_2(s) \cap \mathsf{Pred}(G_2, v_2)| \tag{10}$$

3   Pruning by Considering Paths Compatibility

The motivation for the cited pruning technique and for the new one is as follows. Let's assume that we are checking for an addition of the pair (v_1, v_2) to the current mapping. Now, if the mapping is going to be completed, the local structure of connections around $v_1 \in G_1$ will have to be mapped to a similar local structure around $v_2 \in G_2$. The tests in VF2 considered counts of successor or predecessor *nodes* with different connectivity, we decided to explore checks dedicated to counting *paths* of different kinds. In particular, if there is a path in G_1 of length d starting from vertex v_1, the same path has to be found in G_2 starting from vertex v_2. If such path in G_2 does not exist we can safely omit considering (v_1, v_2) and therefore we can prune the part of the search tree arising from this novel association.

Let us call this general principle "local-paths-based pruning". The realization considered in the present work is based on counting paths in the *underlying* (*undirected*) graph UG corresponding to the original graph. Edge (u, v) is present in the undirected graph if and only if arc (u, v), arc (v, u) or both are present in the original graph. Given a path in UG, it is labeled according to the direction of the arcs in the original graph G. For example, see Fig. 4 for the illustration of a path of kind "out-in-out" arising from v_1. Let us note that we consider *all* paths, including also non-simple ones, with cycles and repeated vertices.

Before starting the algorithm, a pre-processing phase counts the number of paths of length up to d of the different kinds explained above originating at the different vertices of the two graphs G_1 and G_2. When the algorithm encounters a pair of vertices (v_1, v_2) to be tested for possible inclusion in the mapping,

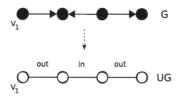

Fig. 4. Path originating from vertex v_1 as considered in BM1

```
1.  COMPATIBLEPATHS (v1, v2, d)
2.    foreach r in 1, . . . , d do
3.      foreach i in 1, . . . , 2^d do
4.        if (PATHSDS[v1][r][i] > PATHSDS[v2][r][i]) then
5.          return false;
6.      done
7.    done
8.    return true;
```

Fig. 5. Pseudo-code for path-counting pruning used in the BM1

one tests whether the number of paths originating at v_1 and v_2 are compatible. In detail, for each length from 1 to d, if the number of paths of at least one kind originating from v_1 is bigger that the number of paths of the same kind originating from v_2, the test is immediately terminated in a negative way. No possible isomorphism can be found by adding (v_1, v_2) to the current mapping.

The new pruning technique presented in this work, hereinafter referred as *BM1* (BATTITIMASCIA-1), is a compatibility check applied before the VF2 check with the aim of further reducing the size of the search tree. Of course the reduction in the number of states visited comes at the cost of an increased complexity of the extra check, and the length d of the paths impacts the precision as well as the cost of the check.

3.1   Data Structures and Computational Complexity

The BM1 algorithm requires an appropriate value of the d parameter. Larger values of d will prune more but at the cost of an increasing computation and memory requirement. It is therefore of interest to evaluate the effectiveness in the reduction of tree size, and the space and time costs as a function of the parameter d. Fig. 5 shows the COMPATIBLEPATHS pseudo-code. The numbers of paths of different length and different kinds are compared and the test returns immediately as soon as, for a specific length and kind, the number of paths in G_2 is less than the number of equivalent paths in G_1.

Fig. 6 shows how the neighborhood information is stored in an ad-hoc data-structure which is computed statically before the actual search takes place. The tree rooted at the vertex whose neighborhood has to be checked shows the information stored in the data structure, i.e., the number of paths of length d labeled with the corresponding "in", "out" arc labels on the edges of the graph.

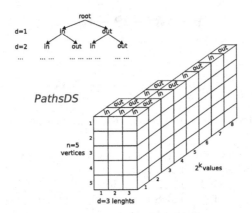

Fig. 6. PATHSDS: data structure storing the number of paths of different kinds originating from a node

During the run, each of the d-length paths checks are performed by making a number of comparisons equal to the leaves in the binary tree representing the neighborhood at the given length. The time complexity of a check for a single pair of vertices is in the worst case equal to:

$$\sum_{i=1}^{d} 2^i = 2^{d+1} - 2$$

The data structure is constructed in a recursive way: the nodes and all their neighbors are visited until all paths of depth d are reached, and during the tail of the recursion all degrees are summed up to fill in the elements.

The time complexity for building the data structure is bounded by $O(n^d)$ and the space occupied by the table is $n * (2^{d+1} - 2)$, see Fig. 6.

4   Computational Experiments for VF2 and BM1

The technique has been tested against chosen instances of the AMALFI Graph DataBase[7]. In order to study the effectiveness of BM1, ten random graphs classes have been selected, having different number of nodes, density, and subgraph sizes.

Each class, which contains 100 instances of the problem, is identified by the size of the sub-graph (si2 means that the number of vertices of the sub-graph is 20% of the graph), the number of vertices of the graph, and the probability η of connection between the vertices. More in detail, the graph is constructed by connecting the vertices with a number of arcs equal to $\eta \cdot |V| \cdot (|V| - 1)$ and by successively adding arcs until the graph is connected [7].

The sub-graph isomorphism between each pair has been tested by means of the original VF2 method, and of the BM1 proposal with values of the parameter d ranging from 1 to 3. For these values of d the initialization time to build the

Table 1. Average states visited by each algorithm for selected graph classes

Instances	VF2	BM1 ($d=1$)	BM1 ($d=2$)	BM1 ($d=3$)
si2_200_01	44 822.60	44 814.14	44 814.14	44 814.14
si2_400_01	505 473.15	505 473.15	505 473.15	505 473.15
si6_200_01	2 524.05	2 100.32	2 100.32	2 100.32
si6_400_01	52 714.16	48 045.39	48 045.39	48 045.39
si6_800_01	5 012 505.53	4 931 432.31	4 931 432.31	4 931 432.31
si2_200_001	428 800.24	155 074.65	118 939.67	115 336.62
si2_400_001	2 315 104.50	838 758.49	818 768.34	818 290.73
si6_200_001	8 756.76	1 292.44	813.93	741.97
si6_400_001	1 303 904.85	82 158.56	47 317.74	46 621.02
si6_800_001	6 006 050.24	1 088 435.49	1 066 752.11	1 066 752.11

Table 2. Average time in μ-seconds spent by each algorithm for selected graph classes

Instances	VF2	BM1 ($d=1$)	BM1 ($d=2$)	BM1 ($d=3$)
si2_200_01	447 500.00	474 900.00	515 200.00	559 600.00
si2_400_01	11 838 800.00	12 612 900.00	13 855 200.00	14 822 800.00
si6_200_01	27 400.00	22 800.00	25 100.00	24 800.00
si6_400_01	1 709 500.00	1 613 200.00	1 733 300.00	1 847 100.00
si6_800_01	375 137 400.00	410 477 100.00	386 953 600.00	461 466 400.00
si2_200_001	2 403 500.00	983 000.00	836 500.00	903 800.00
si2_400_001	18 645 500.00	7 783 200.00	8 501 600.00	9 212 800.00
si6_200_001	45 800.00	5 700.00	3 400.00	4 100.00
si6_400_001	11 379 700.00	660 200.00	401 600.00	411 200.00
si6_800_001	79 632 000.00	13 441 000.00	14 065 400.00	13 944 800.00

Table 3. Steps and time *ratio* between the BM1 algorithm with three different path lengths and VF2. The best length d of the check for the given instance is highlighted. If there is no such value in the row, then VF2 is a better choice.

Instances	BM1 ($d=1$)		BM1 ($d=2$)		BM1 ($d=3$)	
	steps	μ-sec	steps	μ-sec	steps	μ-sec
si2_200_01	1.00	1.06	1.00	1.15	1.00	1.25
si2_400_01	1.00	1.07	1.00	1.17	1.00	1.25
si6_200_01	**0.83**	**0.83**	0.83	0.92	0.83	0.91
si6_400_01	**0.91**	**0.94**	0.91	1.01	0.91	1.08
si6_800_01	0.98	1.09	0.98	1.03	0.98	1.23
si2_200_001	0.36	0.41	**0.28**	**0.35**	0.27	0.38
si2_400_001	**0.36**	**0.42**	0.35	0.46	0.35	0.49
si6_200_001	0.15	0.12	**0.09**	**0.07**	0.08	0.09
si6_400_001	0.06	0.06	**0.04**	**0.04**	**0.04**	**0.04**
si6_800_001	**0.18**	**0.17**	0.18	0.18	0.18	0.18

data structure used by the COMPATIBLEPATHS routine is hardly measurable and not significant w.r.t. the CPU time spent during the tree search. In any case, the total CPU time including initialization is measured in the experiments.

The CPU time spent by the algorithms is measured on our reference machine, having one Xeon processor at 3.4 GHz and 6 GB RAM. The operating system is a Debian GNU/Linux 3.0 with kernel 2.6.15-26-686-smp. All the algorithms are compiled with the g++ compiler with "-O3 -mcpu=pentium4".

To compare the performance we consider both the number of visited states (number of tree nodes) and the CPU time of the different alternatives. For convenience we also report ratios of the above values.

Table 1 summarizes the average number of visited states of the three different BM1(d) compared with VF2. It can be observed that the new pruning technique delivers comparable results on the denser random graphs with $\eta = 0.01$, while it delivers a significantly smaller number of visited states for the sparse graphs with $\eta = 0.001$. For example, for the less dense classes (e.g. si6_200_001) the reduction of visited states reaches 92%. While the additional cut in visited states is negligible for the denser graphs, the cut tends to increase as a function of the d parameter for the sparser graphs. Nonetheless, the speed of reduction in the number of states decreases rapidly as soon as d reaches large values. This can be observed from the Table by considering the reduction when passing from $d = 1$ to $d = 2$ and the much smaller reduction when passing from $d = 2$ to $d = 3$.

Both results, larger effectiveness for sparser graphs and diminishing additional cuts for large d values, are not unexpected. One has to consider that the number of possible paths increases very rapidly as a function of d, in particular if the graph is dense. Because more nodes and edges are available in G_2 to build possible paths one has to expect that the number of paths in G_2 when d increases will become so large that the inequalities in the tests in the COMPATIBLEPATHS routine will be easily satisfied. In practice this means that the bigger cuts are for very small values of d, a positive note when one consider the CPU time spent during the checks.

Table 2 compares the average time spent by the algorithms for finding the mapping between the instances of the different graph classes. For denser graphs ($\eta = 0.01$) the reduction in the number of visited states is too small to see a reduction in the CPU time. For example, the time needed for solving si2_200_001 instances increases with the parameter d because the additional cost of the path compatibility check is not balanced by the reduction in the number of visited states. In the case of sparser graphs ($\eta = 0.001$) BM1 is able to prune the search space more effectively, and the average time for solving the instances decreases with the length of the paths checked, growing again when the increased length does not result in further pruning.

Finally Table 3 summarizes the ratios between the average number of steps and times spent by the algorithms for solving the problem instances. The ratio is between BM1 and VF2, therefore values smaller than 1 implies that BM1 is the winning algorithm.

5   Cumulative Distribution Functions of Randomized Versions

The time spent by the exact algorithm depends on the particular instance of the class of random graphs, but, for each single instance, also on the order in which the vertices are visited. In the original algorithm [4] the choice of the candidate vertices is deterministic. All considered algorithms have been randomized by randomly permuting the vertices in the input graph G_1 before starting. Therefore, in case of ties when considering the next nodes to be mapped, different nodes will be selected in different runs, leading to different results.

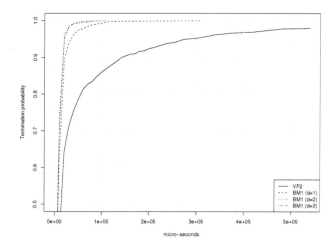

Fig. 7. Probability for randomized version of the algorithm to solve a si6_r001_m200 instance within a fixed time

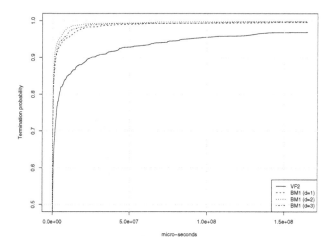

Fig. 8. Probability for randomized version of the algorithm to solve a si6_r001_m400 instance within a fixed time

After randomization, the information of interest about the performance is summarized in the empirical cumulative distribution functions (CDF for short). Fig. 7 and 8 show the probability of terminating within a given amount of microseconds for the VF2 and BM1 algorithms on two selected instances from the sparse graphs. Both algorithms were tested 1000 times with different random seeds on a single representative instance of the si6_r001_m200 and si6_r001_m400 random classes.

6   Algorithm Portfolios

The algorithm portfolios method, first proposed in [8], follows the standard practice in economics to obtain different return-risk profiles in the stock market by combining stocks characterized by individual return-risk values. Risk is related to the standard deviation of return. An evaluation of the portfolio approach on distributions of hard combinatorial search problems is considered for example in [9].

The basic algorithm portfolio method consists of running more algorithms concurrently on a sequential computer, in a time-sharing manner, by allocating a fraction of the total CPU cycles to each of them. The first algorithm to finish determines the termination time of the portfolio, the other algorithms are stopped immediately after one reports the solution.

It is intuitive that the CPU time can be radically reduced in this manner for some statistical distributions of run-times. To clarify ideas, let us consider an extreme example where, depending on the initial random seed, the termination time can be of 1 second or of 1000 seconds, with the same probability. If we run a single process, the expected termination time is approximately of 500 seconds. If we run more copies, the probability that at least one of them is lucky (i.e., that it terminates in 1 second) increases very rapidly towards one. Even if termination is now longer than 1 second because more copies share the same CPU, it is intuitive that the expected time will be much shorter than 500.

The solution time of the portfolio t is related to the one of the individual instances of the algorithm. For a two instance-portfolio it corresponds to:

$$t = \min\{t_1 * 2, t_2 * 2\} \tag{11}$$

where t_1 and t_2 are the time spent by the to running instances to find the solution.

For a portfolio of N component instances, the probability that all instances terminate after t, because of the independence assumption and the slow-down effect, is equal to:

$$(1 - CDF(t/N))^N$$

The probability of the complementary event that at least one terminates before t, and therefore that the portfolio converges before t, is therefore:

$$CDF_{portfolio}(t) = 1 - (1 - CDF(t/N))^N$$

After taking differences one derives the distribution $p(t)$ of the portfolio finishing at time t, from which the expected value $E(t)$ and standard deviation $\sigma = \sqrt{\text{Var}(t)}$ can estimated.

7   Computational Experiments for Portfolios

Considering the si6_r001_m400 instance, the average times for BM1 and VF2 are 6 and 83 seconds, respectively, and both cumulative time distribution functions are heavy tailed having a standard deviations of 89 seconds and 20 minutes respectively. After looking at the probability distributions, both algorithms are good candidates to be blended in a portfolio.

By combining several instances of BM1 in time-sharing, the probability to spend more than 20 seconds for finding a solution decreases from 0.12 to 0.02 with 2 instances and to 0.002 with only 4 instances. Fig. 9 and 10 show the new CDFs of the two algorithms.

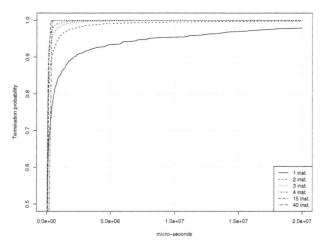

Fig. 9. Probability for a portfolio of several instances of the randomized version of the algorithm to solve a si6_r001_m400 instance within a fixed time

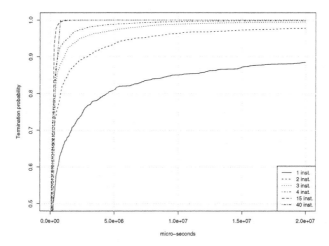

Fig. 10. Probability for a portfolio of several instances of the randomized version of the algorithm to solve a si6_r001_m400 instance within a fixed time

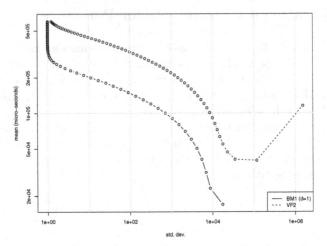

Fig. 11. Portfolios of several instances of the randomized VF2 and BM1 algorithms. The rightmost point in both curves corresponds to a single instance mean and standard deviation, the 2nd point to two instances, the 3rd three and so on. Each point is computed on 1000 runs on si6_r001_m200. Plot is in log-log scale.

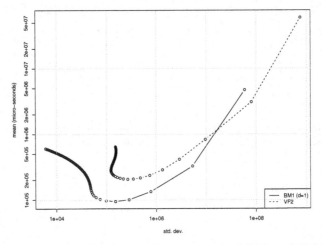

Fig. 12. Portfolios of several instances of the randomized VF2 and BM1 algorithms. The rightmost point in both curves corresponds to a single instance mean and standard deviation, the 2nd point to two instances, the 3rd three and so on. Each point is computed on 1000 runs on si6_r001_m400. Plot is in log-log scale.

The portfolio can be implemented by running one incremental step of each instance of the algorithm at a time, sharing the same process space as well as the path data structure PATHSDS. In this way, the performance degradation is less for the lack of a "real" context switch, and the space as well as the cost of building the shared data structure is shared over the different instances. Fig. 11

and Fig. 12 show the mean solution time versus standard deviation (measured in micro-seconds) for two portfolios algorithms, using VF2 or BM1(1), on a single si6_r001_m200 and si6_r001_m400 instance, respectively. It can be noted how a portfolio consisting of a few copies of the same algorithm rapidly reduced the standard deviation of convergence times. As in the standard portfolio usage, the final choice among Pareto-optimal configurations is then up to the final user, depending on his level of risk-aversion.

8   Conclusions

The novel proposal of this paper consists of the definition of a new parametric pruning technique for the sub-graph isomorphism problem, the analysis of a randomized version of the BM1 and VF2 algorithms, and the study of an algorithm portfolio approach for the problem.

The experimental results on the considered benchmark graphs demonstrate that the proposed pruning technique is effective in reducing the average number of states visited by the BM1 algorithm for sparse random graphs. The reduction in the number of steps and also in the average time spent by the algorithms reaches 92% for some instances. On denser graph classes the reduction in the visited states is not sufficient in order to achieve also a reduction in the average CPU time.

When portfolios are considered, the heavy tails of the empirical run-time distributions of the algorithms can easily be cured by running more randomized instances concurrently on the same machine. Portfolios of algorithms using the proposed pruning technique dominate VF2 on sparse random instances.

Acknowledgement

We acknowledge useful discussions with Krishnam Raju during his internship at Trento about various graph isomorphism heuristics.

References

1. Ullmann, J.: An Algorithm for Subgraph Isomorphism. Journal of the ACM (JACM) 23(1), 31–42 (1976)
2. Bunke, H., Messmer, B.T.: Recent advances in graph matching. IJPRAI 11(1), 169–203 (1997)
3. Larrosa, J., Valiente, G.: Constraint satisfaction algorithms for graph pattern matching. Mathematical Structures in Computer Science 12(4), 403–422 (2002)
4. Cordella, L.P., Pasquale Foggia, C.S., Vento, M.: A (sub)graph isomorphism algorithm for matching large graphs. IEEE Transactions on Pattern Analysis and Machine Intelligence 16(10), 1367–1372 (2004)
5. Garey, M.R., Johnson, D.S.: Computers and Intractability; A Guide to the Theory of NP-Completeness. W. H. Freeman & Co, New York, USA (1990)

6. McKay, B.: Practical graph isomorphism. In: Numerical mathematics and computing, Proc. 10th Manitoba Conf. Winnipeg/Manitoba, pp. 45–87 (1980)
7. http://amalfi.dis.unina.it/
8. Huberman, B.A., Lukose, R.M., Hogg, T.: An economics approach to hard computational problems. Science 275, 51–54 (1997)
9. Gomes, C.P., Selman, B.: Algorithm portfolios. Artif. Intell. 126(1-2), 43–62 (2001)

A Path Relinking Approach
for the Multi-Resource
Generalized Quadratic Assignment Problem

Mutsunori Yagiura[1], Akira Komiya[2], Kenya Kojima[2], Koji Nonobe[3],
Hiroshi Nagamochi[2], Toshihide Ibaraki[4], and Fred Glover[5]

[1] Graduate School of Information Science, Nagoya University, Nagoya, Japan
[2] Graduate School of Informatics, Kyoto University, Kyoto, Japan
[3] Faculty of Engineering and Design, Hosei University, Tokyo, Japan
[4] School of Science and Technology, Kwansei Gakuin University, Sanda, Japan
[5] Leeds School of Business, University of Colorado, Boulder, CO, USA
yagiura@nagoya-u.jp, nag@i.kyoto-u.ac.jp, nonobe@hosei.ac.jp,
ibaraki@ksc.kwansei.ac.jp , fred.glover@colorado.edu

Abstract. We consider the multi-resource generalized quadratic assignment problem (MR-GQAP), which has many applications in various
fields such as production scheduling, and constitutes a natural generalization of the generalized quadratic assignment problem (GQAP) and
the multi-resource generalized assignment problem (MRGAP). We propose a new algorithm PR-CS for this problem that proves highly effective. PR-CS features a path relinking approach, which is a mechanism
for generating new solutions by combining two or more reference solutions. It also features an ejection chain approach, which is embedded in a
neighborhood construction to create more complex and powerful moves.
Computational comparisons on benchmark instances show that PR-CS
is more effective than existing algorithms for GQAP, and is competitive
with existing methods for MRGAP, demonstrating the power of PR-CS
for handling these special instances of MR-GQAP without incorporating
special tailoring to exploit these instances.

1 Introduction

We consider the *multi-resource generalized quadratic assignment problem* (MR-
GQAP), which is a natural generalization of the *generalized quadratic assignment
problem* (GQAP) [1,8] and the *multi-resource generalized assignment problem*
(MRGAP) [2,13]. For this problem, we are given n jobs, m agents, assignment
costs of jobs, a cost matrix between jobs, a cost matrix between agents, and
coefficients for resource constraints. The objective of MR-GQAP is to find a
minimum cost assignment of jobs to agents subject to cardinality constraints
and multi-resource constraints for each agent, where the following two types of
costs are considered: One is individual cost associated with each assignment,
and the other is mutual cost associated with a pair of assignments. MR-GQAP
is NP-hard because both MRGAP and GQAP are NP-hard. MR-GQAP is very

T. Stützle, M. Birattari, and H.H. Hoos (Eds.): SLS 2007, LNCS 4638, pp. 121–135, 2007.
© Springer-Verlag Berlin Heidelberg 2007

general and includes such problems as the graph coloring problem, a special case of the channel assignment problem, and so forth. MR-GQAP is also motivated by some problems emerging from real-world applications such as production scheduling problems in steel industry. It also includes the *quadratic assignment problem* (QAP) as a special case of GQAP, and the *generalized assignment problem* (GAP) as a special case of MRGAP.

For GQAP, Lee and Ma [8] proposed linearization approaches and a branch-and-bound algorithm, and Cordeau et al. [1] have recently proposed a sophisticated memetic algorithm. For MRGAP, Gavish and Pirkul [2] proposed a branch-and-bound algorithm and two simple Lagrangian heuristics, and Yagiura et al. [13] devised a very large-scale neighborhood search algorithm. Nonlinear variants are also discussed, e.g., in [11]. For more restricted special cases such as GAP and QAP, much effort has been devoted to develop efficient exact and heuristic algorithms. To the best of our knowledge, however, not much has been done for GQAP and MRGAP in spite of their practical importance.

In this paper, we propose a heuristic algorithm *PR-CS* (path relinking with chained shift neighborhood) for MR-GQAP. PR-CS features a *path relinking* approach, which provides an *evolutionary* mechanism for generating new solutions by combining two or more reference solutions. The idea of path relinking was proposed by Glover [3,4], and some of its basic aspects were also introduced in an earlier paper by Ibaraki et al. [6]. For more about the general principles of the path relinking approach, see e.g., [7,9]. PR-CS also features the idea of *ejection chains* [5], which is embedded in a neighborhood construction to create more complex and powerful moves. We call the resulting neighborhood the *chained shift neighborhood*, which generalizes standard shift and swap neighborhoods. The problem of judging the existence of a feasible solution for MR-GQAP is NP-complete. We therefore allow our search to visit infeasible solutions that may violate resource constraints as well, and evaluate the amount of the violation as penalty. The performance of the algorithm crucially depends on penalty weights, and hence we incorporate an adaptive mechanism for controlling them to maintain a balance between visiting feasible and infeasible regions.

We conduct computational experiments to observe the effectiveness of each component of the above mentioned methodologies, and to confirm that their combination provides a successful framework for algorithm design. We test PR-CS on MR-GQAP instances generated by us, and on benchmark instances of GQAP and MRGAP. We first compare PR-CS with a basic algorithm (without path relinking mechanism) using only the shift and swap neighborhoods, and observe the effectiveness of the path relinking approach and the chained shift neighborhood. We then compare PR-CS with existing algorithms for GQAP and MRGAP, which are specially tailored to these specific problems, and a general solver for the constraint satisfaction problem. The computational results of PR-CS are quite promising considering its generality.

2   Formulation

Given n jobs $J = \{1, 2, \ldots, n\}$ and m agents $I = \{1, 2, \ldots, m\}$, we undertake to determine a minimum cost assignment of each job to exactly one agent under cardinality constraints and multi-resource constraints for each agent, where s resources $K = \{1, 2, \ldots, s\}$ are considered. In this problem, for $i, i' \in I$, $j, j' \in J$, and $k \in K$, the following data are given as the input:

c_{ij}: the cost of processing job j at agent i,
$u_{jj'}$: the cost coefficient between jobs j and j',
$w_{ii'}$: the cost coefficient between agents i and i',
a_{ijk}: the amount of resource k consumed by job j if it is assigned to agent i,
b_{ik}: the upper bound of resource k available at agent i,
t_i^{UB}: the upper bound on the number of jobs assigned to agent i,
t_i^{LB}: the lower bound on the number of jobs assigned to agent i.

Assigning a job j to an agent i incurs a cost of c_{ij} and consumes an amount a_{ijk} of each resource $k \in K$, whereas the total amount of the resource k available at agent i is b_{ik}. Moreover, for any pair of jobs j, j', assigning jobs j and j' to agents i and i' respectively incurs a cost of $f(u_{jj'}, w_{ii'})$, where $f : R^2 \to R$ is a given function. Throughout the paper, we assume $a_{ijk} \geq 0$ and $b_{ik} > 0$ for all $i \in I$, $j \in J$ and $k \in K$. An assignment is a mapping $\sigma: J \to I$, where $\sigma(j) = i$ means that job j is assigned to agent i. Let

$$J_i^\sigma = \{j \in J \mid \sigma(j) = i\}, \quad \forall i \in I,$$

which is the set of jobs assigned to agent i in assignment σ. Then the problem we consider in this paper is formally described as follows:

$$\text{minimize} \quad cost(\sigma) = \sum_{j \in J} c_{\sigma(j), j} + \sum_{j, j' \in J} f\left(u_{jj'}, w_{\sigma(j)\sigma(j')}\right) \tag{1}$$

$$\text{subject to} \quad t_i^{\text{LB}} \leq |J_i^\sigma| \leq t_i^{\text{UB}}, \qquad \forall i \in I \tag{2}$$

$$\sum_{j \in J_i^\sigma} a_{ijk} \leq b_{ik}, \qquad \forall i \in I \text{ and } \forall k \in K. \tag{3}$$

We mainly consider the case with $f(u, w) = uw$, and call the problem the *multi-resource generalized quadratic assignment problem* (MR-GQAP). We call (2) the *cardinality constraints* and (3) the *resource constraints*. This problem includes GQAP as its special case with $f(u, w) = uw$, $s = 1$, $t_i^{\text{LB}} = 0, t_i^{\text{UB}} = n$ for all $i \in I$ and $a_{ij1} = a_{i'j1}$ for all $i, i' \in I$ and $j \in J$. MRGAP is a special case of MR-GQAP with $f(u, w) \equiv 0$ and $t_i^{\text{LB}} = 0, t_i^{\text{UB}} = n$ for all $i \in I$. QAP is a special case of GQAP with $a_{ij1} = 1$ for all i and j (hence resource constraints can also be described as cardinality constraints), and GAP is a special case of MRGAP with $s = 1$. Note that GQAP does not include GAP because the resource constraint of GQAP must satisfy $a_{ij1} = a_{i'j1}$ for all $i, i' \in I$ and $j \in J$. MR-GQAP is NP-hard in the strong sense, and the (supposedly) simpler problem of judging the existence of a feasible solution for GAP is NP-complete, since the partition problem can be reduced to GAP with $m = 2$.

3 Algorithm

Our algorithm PR-CS is based on local search, where the initial solutions of local search are generated by path relinking. We describe its basic components in the following subsections.

3.1 Local Search, Search Space and Neighborhood

Local search starts from an initial solution σ, and repeatedly replaces the current solution σ with a better solution σ' in its *neighborhood* $N(\sigma)$ until no better solution is found in the neighborhood. The resulting solution is called *locally optimal*. Shift and swap neighborhoods, denoted N_{shift} and N_{swap} respectively, are often used in local search methods for assignment type problems, where $N_{\text{shift}}(\sigma)$ is the set of solutions obtainable from σ by changing the assignment of one job, and $N_{\text{swap}}(\sigma)$ is the set of solutions obtainable from σ by exchanging the assignments of two jobs. The sizes of these neighborhoods are $O(mn)$ and $O(n^2)$, respectively. In addition to these standard neighborhoods, our algorithm uses a *chained shift* neighborhood, which consists of solutions obtainable by certain sequences of shift moves. The chained shift neighborhood $N_{\text{chain}}(\sigma)$ is the set of solutions σ' obtainable from σ by changing the assignments of l ($l = 2, 3, \ldots, n$) arbitrary jobs $j_1, j_2, \ldots, j_l$ simultaneously so that

$$\sigma'(j_r) = \sigma(j_{r-1}), \quad r = 2, 3, \ldots, l$$
$$\sigma'(j_1) = \sigma(j_l).$$

In other words, for $r = 2, 3, \ldots, l$, job j_r is shifted from agent $\sigma(j_r)$ to agent $\sigma(j_{r-1})$ after ejecting job j_{r-1}, and then the cycle is closed by assigning job j_1 to agent $\sigma(j_l)$. The *length* of a chained shift move is the number l of jobs shifted in the move. This is based on the idea of ejection chains by Glover [5]. Since the size of such a neighborhood can become exponential in l, we carefully limit its size by utilizing ejection trees to be explained in Section 3.3. Since $|N_{\text{shift}}| \leq |N_{\text{swap}}| \leq |N_{\text{chain}}|$ usually holds, N_{swap} is searched only if N_{shift} does not contain an improving solution, and N_{chain} is searched only if $N_{\text{shift}} \cup N_{\text{swap}}$ does not contain an improving solution.

The search space of our local search is the set of assignments σ that satisfy the cardinality constraints (2), but may violate the resource constraints (3). Note that it is easy to judge the existence of an assignment σ that satisfies (2): There exists a σ satisfying (2) if and only if

$$\sum_{i \in I} t_i^{\text{LB}} \leq n \leq \sum_{i \in I} t_i^{\text{UB}}. \tag{4}$$

In the rest of this paper, we assume (4) without loss of generality. In the shift neighborhood, we only evaluate solutions satisfying (2). Note that all solutions in the swap and chained shift neighborhoods satisfy (2) if the current solution satisfies (2). The search space is connected if both shift and swap neighborhoods are used; i.e., for any two solutions σ and σ' that satisfy (2), there exists a sequence $\sigma = \sigma_0, \sigma_1, \ldots, \sigma_{l'} = \sigma'$ such that σ_r satisfies (2) and $\sigma_r \in N_{\text{shift}}(\sigma_{r-1}) \cup N_{\text{swap}}(\sigma_{r-1})$ holds for all $r = 1, 2, \ldots, l'$. As the search may visit the infeasible region, we evaluate solutions by an objective function penalized by infeasibility:

$$pcost(\sigma) = cost(\sigma) + \sum_{\substack{i \in I \\ k \in K}} \alpha_{ik} p_{ik}(J_i^\sigma), \tag{5}$$

where

$$p_{ik}(S) = \max \left\{ 0, \sum_{j \in S} a_{ijk} - b_{ik} \right\}$$

for $i \in I$, $k \in K$ and a subset $S \subseteq J$ of the jobs. The parameters α_{ik} (> 0) are adaptively controlled during the search by using the rules similar to those in [12]. The basic idea is simple and intuitively explained as follows: The weights are updated whenever a locally optimal solution is found, and are increased slightly if no feasible solution is found during the last call to local search, and are decreased otherwise (i.e., at least one feasible solution is found during the search). We omit the details due to space limitation.

For convenience, we denote by LS-SS(σ, σ_{incum}) the local search with the shift and swap neighborhoods that starts from a solution σ, where it improves the solution σ to a locally optimal solution and also updates the incumbent solution σ_{incum} (i.e., the best feasible solution found by then) if it finds a better feasible solution during the search.

3.2   An Efficient Implementation of Neighborhood Search

As it takes $O(n^2 + ns + ms)$ time to calculate $pcost$ (5) of one solution from scratch, it takes $O(mn^3 + mn^2s + m^2ns)$ time to calculate all the solutions in the shift neighborhood, if we adopt a naive implementation. In this section, we propose an efficient implementation of the shift neighborhood in which it memorizes the changes of $pcost$ induced by all shift operations in a table of size $O(mn)$. Below, we consider the cost changes and the resulting penalty incurred by shifting job j from agent $\sigma(j)$ to agent i. Let δ_j^{c-}, δ_{ij}^{c+}, δ_j^{p-}, δ_{ij}^{p+} be defined as follows:

$$\delta_j^{c-} = - \sum_{j' \in J \setminus \{j\}} \left\{ f(u_{jj'}, w_{\sigma(j),\sigma(j')}) + f(u_{j'j}, w_{\sigma(j'),\sigma(j)}) \right\}$$
$$- c_{\sigma(j),j} - f(u_{jj}, w_{\sigma(j),\sigma(j)}), \tag{6}$$

$$\delta_{ij}^{c+} = \sum_{j' \in J \setminus \{j\}} \left\{ f(u_{jj'}, w_{i,\sigma(j')}) + f(u_{j'j}, w_{\sigma(j'),i}) \right\} + c_{ij} + f(u_{jj}, w_{ii}), \tag{7}$$

$$\delta_j^{p-} = - \sum_{k \in K} \alpha_{\sigma(j),k} \left\{ p_{\sigma(j),k}(J_{\sigma(j)}^\sigma) - p_{\sigma(j),k}(J_{\sigma(j)}^\sigma \setminus \{j\}) \right\}, \tag{8}$$

$$\delta_{ij}^{p+} = \sum_{k \in K} \alpha_{ik} \{ p_{ik}(J_i^\sigma \cup \{j\}) - p_{ik}(J_i^\sigma) \}. \tag{9}$$

We can decompose the operation of shifting a job j from agent $\sigma(j)$ to agent i into two steps; we first remove job j from agent $\sigma(j)$ and insert it into agent i. In this process, δ_j^{c-} and δ_j^{p-} represent the increases (actually, the decreases times -1) of cost and penalty by the removal of job j from agent $\sigma(j)$, and δ_{ij}^{c+} and δ_{ij}^{p+} represent the increases in the cost and penalty, respectively, by the insertion of job j to agent i. We can calculate the values of δ_j^{c-} and δ_{ij}^{c+} in $O(n)$ time. If the amount of resource $k \in K$ used by agent $i \in I$ (i.e., $\sum_{j \in J_i^\sigma} a_{ijk}$) at the current solution σ is memorized in a table (this table can be prepared in $O((m+n)s)$ time), then δ_j^{p-} and δ_{ij}^{p+} can be calculated in $O(s)$ time. It therefore takes $O(mn(n + s))$ time to compute the table of $\delta_j^{c-}, \delta_{ij}^{c+}, \delta_j^{p-}$, and δ_{ij}^{p+} for all i and j.

If the above table is given, the increase in *pcost* by shifting job j from agent $\sigma(j)$ to agent i is given by

$$\delta_{ij} = \delta_j^{c-} + \delta_j^{p-} + \delta_{ij}^{c+} + \delta_{ij}^{p+} \tag{10}$$

($\delta_{ij} < 0$ means that we can get an improved solution by this shift operation), which can be calculated in $O(1)$ time. We can therefore calculate *pcost* of all the solutions in the shift neighborhood in $O(mn)$ time excluding the time to prepare the table.

We then consider the computation time to renew the table when the current solution is changed by a shift operation. Assume that job j' is shifted from agent i' to agent i'', and let $\hat{\delta}_j^{c-}, \hat{\delta}_{ij}^{c+}, \hat{\delta}_j^{p-}, \hat{\delta}_{ij}^{p+}$ denote the values of $\delta_j^{c-}, \delta_{ij}^{c+}, \delta_j^{p-}, \delta_{ij}^{p+}$ before the shift operation, respectively. For $j \neq j'$, δ_j^{c-} and δ_{ij}^{c+} after the shift move are given by

$$\delta_j^{c-} = \hat{\delta}_j^{c-} - f(u_{jj'}, w_{\sigma(j),i''}) - f(u_{j'j}, w_{i'',\sigma(j)})$$
$$+ f(u_{jj'}, w_{\sigma(j),i'}) + f(u_{j'j}, w_{i',\sigma(j)})$$
$$\delta_{ij}^{c+} = \hat{\delta}_{ij}^{c+} - f(u_{jj'}, w_{ii'}) - f(u_{j'j}, w_{i'i}) + f(u_{jj'}, w_{ii''}) + f(u_{j'j}, w_{i''i}),$$

and the computation time of this update for each pair of i and j is $O(1)$. For the remaining case (i.e., $j = j'$), we need to calculate $\delta_{j'}^{c-}$ by (6), which takes $O(n)$ time, and $\delta_{ij'}^{c+} = \hat{\delta}_{ij'}^{c+}$ holds for all $i \in I$. Therefore it takes $O(mn)$ time to renew the table of δ_j^{c-} and δ_{ij}^{c+} for all pairs of i and j.

For the table of δ_j^{p-} and δ_{ij}^{p+}, we calculate δ_j^{p-} according to (8) for all jobs $j \in J$ such that $\sigma(j) = i'$ or $\sigma(j) = i''$, and calculate δ_{ij}^{p+} according to (9) for all $j \in J$ and $i \in \{i', i''\}$. In this case, we do not need to renew the table for other i and j (i.e., $\delta_j^{p-} = \hat{\delta}_j^{p-}$ holds if $\sigma(j) \neq i', i''$, and $\delta_{ij}^{p+} = \hat{\delta}_{ij}^{p+}$ holds for all $j \in J$ if $i \neq i', i''$). Since the number of δ_j^{p-} and δ_{ij}^{p+} requiring updates is $O(n)$, and it takes $O(s)$ time for each update, it takes $O(ns)$ time to renew δ_j^{p-} and δ_{ij}^{p+} for all i and j.

The time to renew the table of $\sum_{j \in J_i^\sigma} a_{ijk}$ for all i and k is $O(s)$, because we only need to calculate the changes at agents i' and i''. In total, we can renew the table of $\delta_j^{c-}, \delta_{ij}^{c+}, \delta_j^{p-}$ and δ_{ij}^{p+} for all i and j and that of $\sum_{j \in J_i^\sigma} a_{ijk}$ for all i in $O(n(m + s))$ time.

In conclusion, a shift move can be executed in $O(n(m + s))$ time once the tables are initialized. Note that it takes $O(nm(n + s))$ time to initialize the tables when an initial solution for local search is given or the penalty weights α_{ik} are changed. Although time for initializing the tables is larger than the computation time needed for each move, we can usually ignore it because the number of moves in a local search is much larger than the number of initialization of tables. For many instances, $s \ll m$ holds, and the computation time for a shift move becomes $O(mn)$, which is the same as the size of N_{shift}. In such cases, we can evaluate one solution in the shift neighborhood in $O(1)$ amortized time.

Based on a similar idea, we can evaluate a solution in the swap neighborhood in $O(s)$ time using $\delta_j^{c-}, \delta_{ij}^{c+}, \delta_j^{p-}$ and δ_{ij}^{p+}, and renew the tables in $O(n(m + s))$ time. (The details are omitted due to space limitation.) For many instances, s can be considered as a fixed constant and $m \leq n$ hold, and hence the computation time for a move becomes $O(n^2)$, which is the same as the size of N_{swap}.

3.3 Search in the Chained Shift Neighborhood

In this section, we briefly explain the idea of our algorithm for finding an improved solution in the chained shift neighborhood using ejection trees. The ejection tree is a rooted tree, in which each vertex corresponds to a job, and the path from the root to a vertex corresponds to a chained shift move.

We consider a set of n ejection trees $T(\sigma, 1), T(\sigma, 2), \ldots, T(\sigma, n)$ corresponding to the current solution σ. The root vertex of $T(\sigma, j)$ corresponds to job j, and other vertices correspond to other jobs. Let $j(v)$ denote the job assigned to a vertex v, and $\rho(v)$ denote the parent of v with depth $d_v \geq 1$. Let $j_0^v(= j), j_1^v, \ldots, j_{d_v}^v$ denote the sequence of jobs in the path from the root to a vertex v in depth d_v. Then the chained shift move corresponding to this path is as follows:

$$\sigma'(j_d^v) = \sigma(j_{d-1}^v), \quad d = 1, 2, \ldots, d_v$$
$$\sigma'(j_0^v) = \sigma(j_{d_v}^v),$$

where σ' is the new solution generated by the move. Let σ_v be the solution obtained by the chained shift operation that corresponds to the path to v from its root.

It is clear that we can generate all possible solutions in the chained shift neighborhood by considering appropriate ejection trees; however, generating all solutions in this neighborhood is not realistic. We therefore limit the search by the following heuristic rules.

- The search is restricted to the vertices of depth $\leq d_{\max}$ (a parameter).
- In each depth d, we choose the vertices with the smallest $\lceil \gamma/d \rceil$ (γ is a parameter) values of $\Delta^-(v)$ among the set of vertices generated in depth $d (\geq 1)$, and generate only the descendants of the chosen vertices, where $\Delta^-(v)$ is the difference in *pcost* between the current solution σ and the incomplete solution obtained by ejecting the assignment of the job j_0^v from the solution σ_v.

In the experiment in Section 4, we set $d_{\max} = \min\{m, 5\}$ and $\gamma = 4$.

We implement our algorithm so that it evaluates each solution σ_v in $O(d_v + s)$ time, by using an idea similar to those in Section 3.2; however, its details are quite complicated and are omitted. The whole computation time to search the chained shift neighborhood is $O(n^2 s + nm)$.

Even with such an elaborate implementation, the search in the chained shift neighborhood is still expensive compared to the search in the shift and swap neighborhoods. We therefore invoke the search in the chained shift neighborhood only if the current solution σ is locally optimal with respect to N_{shift} and N_{swap}, and $pcost(\sigma) < 1.01cost(\sigma_{\text{incum}})$ holds, where σ_{incum} is the incumbent solution. We denote by LS-CS$(\sigma, \sigma_{\text{incum}})$ the local search with the shift, swap and chained shift neighborhoods that starts from a solution σ (it receives the current solution σ and the incumbent solution σ_{incum} and modifies them if possible).

3.4   Path Relinking and Reference Set

Path Relinking. We generate initial solutions for LS-SS and/or LS-CS by a path relinking approach, which is a method to construct solutions from two solutions. We define a path to be the set of solutions obtained by repeatedly applying the shift operations from a solution to the other. If two solutions σ_1 and σ_2 are given, the path relinking gives a set of initial solutions S along the path between σ_1 and σ_2. Let J' be the set of jobs assigned to different agents between σ_1 and σ_2. To construct a path from σ_1 to σ_2, in each step, we shift a job $j \in J'$ such that $\delta_{\sigma_2(j),j}$ (i.e., the increase in $pcost$ calculated by (10)) is minimum. In our algorithm, we apply local search to at most ω solutions in the path having small $pcost$, where ω is a parameter. For a given pair of σ_1 and σ_2, our path relinking procedure, denoted PR(σ_1, σ_2), is formally described as follows.

Procedure PR(σ_1, σ_2)
Step 1. Let $\sigma := \sigma_1$, $S := \emptyset$ and $J' := \{j \in J \mid \sigma_1(j) \neq \sigma_2(j)\}$.
Step 2. Choose a job $j \in J'$ with minimum $\delta_{\sigma_2(j),j}$, and let $\sigma(j) := \sigma_2(j)$.
Step 3. Let $S := S \cup \{\sigma\}$, and remove j from J'. If $|J'| \geq 2$, return to Step 2; otherwise proceed to Step 4.
Step 4. If $|S| \leq \omega$, output S, otherwise let S' be the set of solutions in S with ω smallest values of $pcost$, and output S'.

Reference Set. We keep a set R of good solutions, and choose the two solutions σ_1 and σ_2 for path relinking from R. It is preferable to keep good solutions in the reference set R to make path relinking more effective, while similar solutions in R are not desirable from the view point of diversification. As candidates for R, we test only locally optimal solutions obtained in the previous call to local search.

We define the distance D between two solutions σ_1 and σ_2 to be the number of jobs that are assigned to different agents; i.e.,

$$D(\sigma_1, \sigma_2) = |\{j \in J \mid \sigma_1(j) \neq \sigma_2(j)\}|.$$

We keep R in such a way that the distance between any two solutions is at least κ (a parameter) for attaining diversification.

We now explain the rule for renewing the reference set R. Let σ be the locally optimal solution obtained in the previous local search, and σ_{worst} be a solution with the maximum *pcost* in R. We consider the following two cases: (1) All solutions in R have distances from σ larger than or equal to κ, and (2) otherwise. In case (1), if $|R| < \zeta$ (a parameter) holds, then we add σ into R; otherwise, if $pcost(\sigma) < pcost(\sigma_{\text{worst}})$ holds, then we exchange σ and σ_{worst}, i.e., we let $R := R \setminus \{\sigma_{\text{worst}}\} \cup \{\sigma\}$. In case (2), if $pcost(\sigma) < pcost(\sigma')$ holds for all $\sigma' \in R$ such that distance between σ and σ' is smaller than κ, then we add σ into R, and remove all the solutions whose distance from σ is smaller than κ. The procedure to renew the reference set for a given locally optimal solution σ, denoted $\text{RNR}(\sigma, R, \kappa)$, is summarized as follows.

Procedure RNR(σ, R, κ)
Step 1. If $D(\sigma, \sigma') \geq \kappa$ holds for all $\sigma' \in R$, go to Step 2; otherwise go to Step 3.
Step 2. If $|R| < \zeta$, let $\tau := +\infty$ and $A := \emptyset$; otherwise let σ_{worst} be a solution in R such that $pcost(\sigma') \leq pcost(\sigma_{\text{worst}})$ for all $\sigma' \in R$, and then let $\tau := pcost(\sigma_{\text{worst}})$ and $A := \{\sigma_{\text{worst}}\}$. Go to Step 4.
Step 3. Let $A := \{\sigma' \in R \mid D(\sigma, \sigma') < \kappa\}$, and let σ_{best} be a solution in A such that $pcost(\sigma') \geq pcost(\sigma_{\text{best}})$ for all $\sigma' \in A$. Then let $\tau := pcost(\sigma_{\text{best}})$ and go to Step 4.
Step 4. If $pcost(\sigma) < \tau$, then let $R := R \setminus A \cup \{\sigma\}$.

3.5   The Whole Framework of the Algorithm

Our algorithm PR-CS basically applies LS-SS or LS-CS to solutions generated by the path relinking method. Its details are summarized in this section.

At the beginning of the search, the reference set R is empty, and the size of R may increase or decrease when procedure RNR is called. If $|R| < \zeta$ holds and the set of initial solutions generated by the previous call to the path relinking is exhausted, then we apply the local search to randomly generated solutions until $|R| = \zeta$ holds, where the generated locally optimal solutions are added to R according to the rule in Section 3.4.

We also adopt the following rules to realize intensification and diversification. Let R_{best} be the set of solutions in R with ξ smallest values of *pcost*, where ξ is a parameter. Then we choose σ_1 from R_{best} (to intensify the search) and σ_2 from R both randomly. After choosing two solutions, we add random shifts to σ_2 in 1% of jobs for diversification. Moreover, we increase the minimum distance for renewing the reference set to 2κ if the number of calls to local search (LS-SS or LS-CS) after the last update of the incumbent solution is more than or equal to 2θ (θ is a parameter).

As the search in the chained shift neighborhood takes much time compared to shift and swap neighborhoods, we invoke LS-CS only if the current penalty weights are judged as appropriate,[1] and the number r of calls to local search from the last update of the incumbent solution satisfies $\theta \leq r < 2\theta$ or $r \geq 4\theta$. (Recall that we double the parameter κ for procedure RNR for diversification when $r \geq 2\theta$ holds. When this rule applies, we first use LS-SS in its early stage, i.e., when $2\theta \leq r < 4\theta$ holds.)

The whole framework of our algorithm is described as follows, where ω, ζ, ξ and κ are parameters. In the computational experiments in Section 4, we set $\omega = 5$, $\zeta = 10$, $\xi = 3$, $\theta = 2\xi\zeta\omega$, and the initial value of κ to 3. We stop the search when a prespecified amount of time is spent.

Algorithm PR-CS
Phase 1 (Initialization)
Step 1. Let $R := \emptyset$, $S := \emptyset$, $r := 0$ and $\kappa' := \kappa$.
Step 2. Randomly generate a solution that satisfies (2) (recall that this is always possible by the assumption (4)), and apply a local search with the shift and swap neighborhoods, where each solution σ is evaluated by the total penalty excess $\sum_{i \in I, k \in K} p_{ik}(J_i^\sigma)$ breaking ties by $cost(\sigma)$. Let σ be the locally optimal solution obtained by the local search. If σ is feasible, let $\sigma_{\text{incum}} := \sigma$ (σ_{incum} keeps the incumbent solution).
Step 3. Initialize the penalty weights.
Phase 2 (Construction of the reference set)
Step 4. Let σ be a randomly generated solution that satisfies (2). If the current penalty weights are appropriate, $pcost(\sigma) < 1.01cost(\sigma_{\text{incum}})$ holds, and $\theta \leq r < 2\theta$ or $r \geq 4\theta$ holds, invoke LS-CS($\sigma, \sigma_{\text{incum}}$); otherwise invoke LS-SS($\sigma, \sigma_{\text{incum}}$). Let $r := r+1$. If $r = 2\theta$, then let $\kappa' := 2\kappa$. If σ_{incum} is updated, then let $r := 0$ and $\kappa' := \kappa$. Update the penalty weights.
Step 5. Invoke RNR(σ, R, κ'). If the stopping criterion is satisfied, output the best feasible solution σ_{incum} found during the search and halt.
Phase 3 (Construction of the set of initial solutions)
Step 6. If $|R| < \zeta$, go to Step 4; otherwise go to Step 7.
Step 7. Let R_{best} be the subset of R containing the solutions with ξ smallest values of $pcost$. Choose two solutions σ_1 and σ_2 ($\sigma_1 \neq \sigma_2$) randomly, σ_1 from R_{best} and σ_2 from R.
Step 8. Apply random shifts to σ_2, and let the new solution be σ_2'. Invoke PR(σ_1, σ_2') and let S be its output.
Phase 4 (Improvement of solutions)
Step 9. Randomly choose a solution σ in S, and remove it from S. Then, if the current penalty weights are appropriate, $pcost(\sigma) < 1.01cost(\sigma_{\text{incum}})$ holds, and $\theta \leq r < 2\theta$ or $r \geq 4\theta$ holds, invoke LS-CS($\sigma, \sigma_{\text{incum}}$); otherwise invoke LS-SS($\sigma, \sigma_{\text{incum}}$). Let $r := r + 1$. If $r = 2\theta$, then let $\kappa' := 2\kappa$. If σ_{incum} is updated, then let $r := 0$ and $\kappa' := \kappa$. Update the penalty weights.
Step 10. Invoke RNR(σ, R, κ'). If the stopping criterion is satisfied, output the best feasible solution σ_{incum} found during the search and halt.
Step 11. If $S \neq \emptyset$, go to Step 9, otherwise go to Step 6.

---

[1] We judge the current penalty weights to be appropriate if the rule for incrementing the penalty weights and that for decrementing them are both invoked ten times.

4   Computational Experiments

We conducted computational experiments of our algorithm PR-CS for MR-GQAP, and also compared the results with those of existing algorithms for GQAP and MRGAP. PR-CS was coded in C++ language and run on an IBM IntelliStation Z Pro (two Intel Xeon 3.2 GHz processors with 2 GB memory, where the computation was done on a single processor). The instances used in our experiments are available at our site.[2]

4.1   Multi-Resource Generalized Quadratic Assignment Problem

We generated test instances by adding quadratic costs to the benchmark instances of GAP and MRGAP called types C, D and E (see [12,13] for the details of these types). For each instance, we added three different types of quadratic costs, types 1, 2 and 3. For all the three types, we used $f(u,v) = uv$. In type 1, we set $u_{jj} = 0$ for all j and $w_{ii} = 0$ for all i, and we generated $u_{jj'}$ for all $j \neq j'$ and $w_{ii'}$ for all $i \neq i'$ randomly. In type 2, the quadratic cost takes a positive value only if two jobs are assigned to the same agent; i.e., $u_{jj'} = 0$ (1) if $j = j'$ $(j \neq j')$ and $w_{ii'} = C$ (0) if $i = i'$ $(i \neq i')$ (C is a positive constant chosen from $[1, 20]$). Type 3 is the cost that takes a positive value only if two jobs are assigned to different agents; i.e., $u_{jj'} = 0$ (1) if $j = j'$ $(j \neq j')$ and $w_{ii'} = 0$ (C) if $i = i'$ $(i \neq i')$ (C is a positive constant chosen from $[1, 10]$).

To see the effectiveness of path relinking and cyclic neighborhood, we compare our algorithm PR-CS with the random multi-start local search (denoted MLS) that repeatedly calls LS-SS from randomly generated solutions, and the PR-CS algorithm without the chained shift neighborhood (denoted PR-SS). In MLS, we incorporate the adaptive control mechanism of penalty weights, and PR-SS is exactly the same as PR-CS except that it does not invoke LS-CS. We also compare PR-CS with a general solver for the constraint satisfaction problem by Nonobe and Ibaraki (denoted NI)[10]. We also tested CPLEX 9.0.0 (a general mixed integer programming solver); however, it took too much time even to find a feasible solution; e.g., CPLEX could not find a feasible solution for an instance of $n = 100$ and $m = 10$ in one hour on a PC with Xeon 3.01 GHz.

PR-SS and MLS were also coded in C++ language and run on the same PC as PR-CS. NI was run on a PC with Intel Pentium III 1 GHz and 1GB memory. The time limits for MLS, PR-SS and PR-CS are 300 seconds, and that for NI is 1200 seconds. The number of runs of each algorithm for each instance is one.

Table 1 shows the costs obtained by the tested algorithms, where the column "type" shows the type of the original GAP or MRGAP instance, the column "quadratic cost" shows the type of the quadratic cost, and each '∗' mark indicates the best objective value among the four algorithms in the table. For MR-GQAP, the performance of PR-CS and PR-SS is much better than MLS and NI, and that of PR-CS is slightly better than PR-SS.

---

[2] URL of our site: http://www.al.cm.is.nagoya-u.ac.jp/~yagiura/mrgqap/

Table 1. Comparison of four algorithms for instances of MR-GQAP

instance	n	m	s	type	quadratic cost	NI	MLS	PR-SS	PR-CS
qc05501	50	5	1	C	1	15897	*15822	*15822	*15822
qc05502	50	5	1	C	2	*1315	*1315	*1315	*1315
qc05503	50	5	1	C	3	2849	*2846	*2846	*2846
qc101001	100	10	1	C	1	41798	40430	*40320	*40320
qc101002	100	10	1	C	2	23290	23150	*23110	*23110
qc101003	100	10	1	C	3	104950	103840	*103710	*103710
mqc1010041	100	10	4	C	1	24263	20182	*20127	20137
mqc1010042	100	10	4	C	2	10472	10458	*10452	*10452
mqc1010043	100	10	4	C	3	89442	86066	*86060	*86060
qc102001	200	10	1	C	1	220452	214912	214094	*214028
qc102002	200	10	1	C	2	14379	14246	14224	*14222
qc102003	200	10	1	C	3	75658	72690	*72215	*72215
mqc1020041	200	10	4	C	1	62465	50833	50387	*50193
mqc1020042	200	10	4	C	2	14375	14263	14237	*14234
mqc1020043	200	10	4	C	3	76275	72631	72519	*72486
qd05501	50	5	1	D	1	38636	*38543	*38543	*38543
qd05502	50	5	1	D	2	24411	24420	*24309	*24309
qd05503	50	5	1	D	3	52770	52620	*52460	*52460
qd101001	100	10	1	D	1	43019	36686	36571	*36540
qd101002	100	10	1	D	2	8559	8328	8178	*8171
qd101003	100	10	1	D	3	15960	15390	*15048	15159
mqd1010041	100	10	4	D	1	23486	18653	*18593	*18593
mqd1010042	100	10	4	D	2	8617	8414	8231	*8228
mqd1010043	100	10	4	D	3	16077	15526	*15174	*15174
qd102001	200	10	1	D	1	269574	250141	243529	*243234
qd102002	200	10	1	D	2	25012	24036	23885	*23884
qd102003	200	10	1	D	3	84833	82215	79172	*79160
mqd1020041	200	10	4	D	1	60342	44220	*43766	*43766
mqd1020042	200	10	4	D	2	25058	24212	*23896	23897
mqd1020043	200	10	4	D	3	84707	79630	78968	*78962
qe05501	50	5	1	E	1	64434	*64148	*64148	*64148
qe05502	50	5	1	E	2	8691	*8635	*8635	*8635
qe05503	50	5	1	E	3	10320	*10309	*10309	*10309
qe101001	100	10	1	E	1	53054	49992	*49210	*49210
qe101002	100	10	1	E	2	30005	29859	29723	*29720
qe101003	100	10	1	E	3	30955	29866	*29541	29548
mqe1010041	100	10	4	E	1	39774	36354	*35607	*35607
mqe1010042	100	10	4	E	2	30359	30133	29799	*29787
mqe1010043	100	10	4	E	3	31106	30139	29639	*29628
qe102001	200	10	1	E	1	324138	306782	303556	*303468
qe102002	200	10	1	E	2	64797	61426	61342	*61338
qe102003	200	10	1	E	3	146321	133018	*131261	131287
mqe1020041	200	10	4	E	1	89056	79750	*78338	78357
mqe1020042	200	10	4	E	2	64698	61809	61398	*61397
mqe1020043	200	10	4	E	3	148245	132445	*131299	131304

4.2   Generalized Quadratic Assignment Problem

We tested benchmark instances of GQAP [1,8], and compared PR-CS with the
memetic algorithm by Cordeau et al. (denoted MA)[1], which is specially tailored
for GQAP. We refer the results of MA reported in [1], in which MA was run on
a SUN workstation (1.2 GHz).[3]

Table 2. Comparison of MA and PR-CS for instances of GQAP

instance	n	m	MA		PR-CS		
			value	time (s)	value	TTB (s)	TL (s)
20-15-35	20	15	1471896	96	1471896	0.300	9
20-15-55	20	15	1723638	102	1723638	0.204	10
20-15-75	20	15	1953188	102	1953188	4.856	10
30-06-95	30	6	5160920	114	5160920	0.132	11
30-07-75	30	7	4383923	156	4383923	0.056	15
30-08-55	30	8	3501695	96	3501695	0.496	9
30-10-65	30	10	3620959	210	3620959	1.440	21
30-20-35	30	20	3379359	564	3379359	0.528	50
30-20-55	30	20	3593105	462	3593105	0.756	46
30-20-75	30	20	4050938	522	4050938	0.084	50
30-20-95	30	20	5710645	5232	5710645	511.024	520
35-15-35	35	15	4456670	456	4456670	0.348	45
35-15-55	35	15	4639128	384	4639128	0.492	38
35-15-75	35	15	6301723	396	6301723	26.150	39
35-15-95	35	15	6670264	864	6670264	31.326	50
40-07-75	40	7	7405793	180	7405793	0.308	18
40-09-95	40	9	7667719	1140	7667719	9.697	50
40-10-65	40	10	7265559	240	7265559	0.368	24
50-10-65	50	10	10513029	504	10513029	0.324	50
50-10-75	50	10	11217503	606	11217503	0.544	50
50-10-95	50	10	12845598	1254	12845598	0.276	50
CPU			Sun 1.2 GHz		Xeon 3.2 GHz		

For benchmark instances of Lee and Ma [8], both PR-CS and MA succeeded
in obtaining exact optimal solutions for all instances. The computation time of
PR-CS to obtain an optimal solution for each instance is less than 0.2 seconds,
while the computation time of MA ranges from 1 to 8 seconds. These instances
are somewhat easy and CPLEX was able to solve all of them exactly in less than
10 seconds for two-thirds of the instances, in 10–60 seconds for the rest except
one instance, and with more than 200 seconds for one instance.

Table 2 shows the results for benchmark instances of Cordeau et al. [1]. The
column "time" shows the computation time of MA, the column "TL" shows
the time limit of PR-CS, the column "TTB" shows the time when the best

---

[3] According to the SPEC site (http://www.spec.org/), the values of SPECint2000 are
around 700–722 for Sun workstations (1.2 GHz) and around 1289–1579 for Xeon (3.2
GHz). Hence the speed of the Xeon seems to be 2–3 times faster than the Sun.

Table 3. Comparison with NI, TS and MLS for MRGAP instances

type	NI	TS	PR-CS
C	0.140	0.060	0.052
D	2.118	0.885	0.992
E	1.682	0.358	0.464

solutions were found and the columns "value" show the objective values of the best solutions obtained by the algorithms. We set the time limit of PR-CS to the smaller value of one tenth of the computation time of MA and 50 seconds, except for instance 30-20-95 for which we set the time limit to one tenth of the computation time of MA. These time limits are not longer than the time spent by MA if the speed of computers are taken into consideration. From Table 2, we can observe that the solution values obtained by the two algorithms are exactly the same for all instances.

These results indicate that PR-CS is at least as good as MA. The computation times reported for MA are the time when it stopped, and hence it is not easy to draw a decisive conclusion; however, PR-CS seems to spend less computation time than MA.

4.3   Multi-Resource Generalized Assignment Problem

We test algorithm PR-CS on benchmark instances of MRGAP with up to 200 jobs, 20 agents and 8 resources, and compare its performance with NI and the tabu search by Yagiura et al. (denoted TS)[13], which is specially tailored for MRGAP. TS and NI for MRGAP were run on a workstation Sun Ultra 2 Model 2300 (300 MHz, 1 GB memory).[4] The time limits of NI and TS are 300 and 600 seconds for $n = 100$ and 200, respectively, and the time limits of PR-CS are 30 and 60 seconds for $n = 100$ and 200, respectively. The number of runs of each algorithm for each instance is one.

Table 3 shows the average gap in % of the costs obtained by the algorithms within the time limit from the lower bound reported in [13], where the average was taken over 24 instances for each of types C, D and E. From the table, we can observe that the performance of PR-CS is much better than NI, and is competitive with TS. It is worth noting that the average gap of PR-CS is slightly better than TS for type C instances. Considering its generality, these competitive results are quite encouraging.

5   Conclusion

In this paper, we proposed a heuristic algorithm PR-CS for MR-GQAP, which incorporated the path relinking and ejection chain components. Through computational experiments on randomly generated instances of MR-GQAP, we

---

[4] We estimate that Xeon (3.2 GHz) is about 10 times faster than the Sun (300 MHz).

confirmed that such algorithmic components are effective for improving the performance of local search. We also observed that PR-CS is more efficient than general purpose solvers developed for constraint satisfaction and mixed integer programming problems. Computational results on benchmark instances of GQAP and MRGAP, special cases of MR-GQAP, disclosed that PR-CS was highly efficient in that its performance was competitive with (or sometimes even better than) existing algorithms specially tailored for GQAP and MRGAP. Considering the generality of our algorithm PR-CS, these results are quite satisfactory.

References

1. Cordeau, J., Gaudioso, M., Laporte, G., Moccia, L.: A memetic heuristic for the generalized quadratic assignment problem. INFORMS Journal on Computing 18, 433–443 (2006)
2. Gavish, B., Pirkul, H.: Algorithms for the multi-resource generalized assignment problem. Management Science 37, 695–713 (1991)
3. Glover, F.: Genetic algorithms and scatter search: unsuspected potentials. Statistics and Computing 4, 131–140 (1994)
4. Glover, F.: Tabu search for nonlinear and parametric optimization (with links to genetic algorithms. Discrete Applied Mathematics 49, 231–255 (1994)
5. Glover, F.: Ejection chains, reference structures and alternating path methods for traveling salesman problems, Research Report, University of Colorado, Boulder, CO. Discrete Applied Mathematics 65, 223–253 (1996)
6. Ibaraki, T., Ohashi, T., Mine, H.: A heuristic algorithm for mixed-integer programming problems. Mathematical Programming Study 2, 115–136 (1974)
7. Laguna, M., Martí, R.: Scatter Search: Methodology and Implementations in C. Kluwer Academic Publishers, Boston (2003)
8. Lee, C., Ma, Z.: The generalized quadratic assignment problem, Technical Report. Department of Mechanical and Industrial Engineering, University of Toronto, Toronto, Ontario, Canada (2003)
9. Martí, R., Laguna, M., Glover, F.: Principles of scatter search. European Journal of Operational Research 169, 359–372 (2006)
10. Nonobe, K., Ibaraki, T.: A tabu search approach to the CSP (constraint satisfaction problem) as a general problem solver. European Journal of Operational Research 106, 599–623 (1998)
11. Voss, S.: Heuristics for nonlinear assignment problems. In: Pardalos, P.M., Pitsoulis, L.S. (eds.) Nonlinear Assignment Problems, pp. 175–215. Kluwer Academic Publishers, Dordrecht (2000)
12. Yagiura, M., Ibaraki, T., Glover, F.: An ejection chain approach for the generalized assignment problem. INFORMS Journal on Computing 16, 133–151 (2004)
13. Yagiura, M., Iwasaki, S., Ibaraki, T., Glover, F.: A very large-scale neighborhood search algorithm for the multi-resource generalized assignment problem. Discrete Optimization 1, 87–98 (2004)

A Practical Solution Using Simulated Annealing for General Routing Problems with Nodes, Edges, and Arcs

Hisafumi Kokubugata, Ayako Moriyama, and Hironao Kawashima

Department of Administration Engineering, Keio University,
Hiyoshi, Yokohama, Japan
kokubu@mita.cc.keio.ac.jp, a.moriyama@ana.co.jp, kawashima@ae.keio.ac.jp

Abstract. A new practical solution of the general routing problems with nodes, edges, and arcs (NEARP) has been developed. The method is characterized by a primitive data modeling and a simple optimization procedure based on simulated annealing. The data structure of the method, that is traveling routes of a number of vehicles, is expressed as a string. The solutions generated by the proposed method are compared with those of another method by conducting computational experiments on instances of the NEARP. Moreover, it is shown that the proposed method is adaptable to additional conditions.

1 Introduction

About 20% of total CO_2 emissions in Japan are caused by transportation. In particular, over one third of 20% come from freight transportation [1]. In urban areas, demand for high-quality services such as small volume high frequency pickups and deliveries with tight time windows have been imposed by many clients. The rationalization in terms of decreasing the total travel time is aimed not only at reducing operational costs in each carrier but also at relieving traffic congestion, reducing the amount of exhaust fumes and saving energy.

Proposing decision support tools to improve city logistics, we have developed a new practical solution for the general routing problem with nodes, edges, and arcs (NEARP). The vehicle routing problem (VRP) involves the design of a set of minimum cost vehicle trips, originating and ending at a depot, for a fleet of vehicles with loading capacity that services a set of client spots with required demands. In the case where service time windows are imposed by clients, the routing problem is called the VRP with time windows (VRPTW). The VRP belongs to NP-hard problems. Even concerning the simple VRP, exact methods are not fit for large problems. Therefore, heuristics have been important in the application of the VRP. In the last two decades, many solutions using meta-heuristics have been proposed for the VRP. However, most of them incorporate elaborate procedures. We have proposed an original solution for the VRPs. It is characterized by applying a simple algorithm using simulated annealing based on a primitive data structure. We have dealt with a standard VRPTW [2], a

T. Stützle, M. Birattari, and H.H. Hoos (Eds.): SLS 2007, LNCS 4638, pp. 136–149, 2007.

VRPTW with backhauls [3], and a VRPTW with various conditions including repeated trips and multiple depots [4]. Although the method is quite simple, the quality of solutions is as good as the best known solutions previously reported in the literature.

Observing delivery operations in urban areas, it is found that some delivery or pickup demands belong not to spots but to streets. This circumstance includes the case where the demands are densely located along a street such as postal deliveries, and the case where the demand belongs to a street itself such as waste collection and snow removal. In these cases, it is more suitable to formulate as the capacitated arc routing problem (CARP) rather than as the VRP. The CARP consists of determining a set of vehicle trips at minimum total cost, such that each trip starts and ends at a depot, each required undirected edge is serviced by one single trip, and the total demand handled by any vehicle does not exceed its loading capacity. The CARP is also NP-hard and exact methods are not fit to solve large problems. Although the CARP can be converted into an equivalent VRP, the conversion causes a hardly acceptable increase in the size of the derived VRP. Therefore, the direct solutions of the CARP is preferred. In spite of the fact that good solutions using metaheuristics have been proposed for the CARP, most of them incorporate elaborate procedures [5,6]. We also have applied our simple solution method to the CARP [7].

Although the CARP is defined with the intention for it to be applied to practical arc routing operations, the pure CARP is only able to express arc routing operations in the real world imperfectly. To take waste collection as an example, there are some punctual dumps (such as factories, schools, and hospitals) that put out a large amount of waste, while other small waste dumps along a street are considered as the grouped arc demand. Moreover, the CARP handles demands which belong to only undirected arcs. There are many one-way streets in urban areas. Besides, even in two-way streets, vehicles often collect waste along one side of the street only, because broad streets are often split by central reservations. In order to fit the model to the routing situations in the real world, Prins et al. defined a general routing problem with nodes, edges, and arcs (NEARP) that handles demands which belong to any of nodes, (undirected) edges and (directed) arcs [8]. They applied a memetic algorithm, which has already been applied to the CARP in [6], to the NEARP and examined their solution in a set of NEARP instances. Incidentally, another approach to a similar situation was proposed by Oppen et al. [9]. They aggregate the densely located nodes along a street to a super node, while they treat the isolated node as it is. After the aggregation is done, they solve the aggregated VRP using the tabu search. They treat super nodes as well as (undirected) edges. It might be understood that they proposed another practical approach to a routing problem with nodes and edges.

In this paper, we apply our simpler method, which has been designed for the VRP and applied to the CARP, to the NEARP defined by Prins et al. Moreover, we try to compare our solution with the one given by them by applying our solution to the same set of instances.

2   The Node, Edge and Arc Routing Problem (NEARP)

As mentioned in the previous section, Prins et al. defined the NEARP [8]. In this section, we make a brief sketch of it by referencing them. The NEARP allows a mixed network including required nodes, edges and arcs. Two distinct costs are handled for each link: one is the deadheading cost, i.e., the cost for a traversal without service (called deadhead by transporters) and the other is service cost, when the link is traversed to be serviced.

2.1   Description of the NEARP

Prins et al. defined the NEARP and described it in the style used by Crescenzi et al. [10] for their compendium of NP optimization problems. It is quoted as follows.

- INSTANCE: Mixed graph $G = (V, E, A)$, initial vertex $s \in V$, vehicle capacity $W \in \mathbb{N}$, subset $V_R \subseteq V$, subset $E_R \subseteq E$, subset $A_R \subseteq A$, traversal cost $c(u) \in \mathbb{N}$ for each 'entity' $u \in V \cup E \cup A$, demand $q(u) \in \mathbb{N}$ and processing cost $p(u) \in \mathbb{N}$ for each required entity (task) $u \in V_R \cup E_R \cup A_R$.
- SOLUTION: A set of cycles (trips), each containing the initial vertex s, that may traverse each entity any number of times but process each task exactly once. The total demand processed by any trip cannot exceed W.
- MEASURE: The total cost of the trips, to be minimized. The cost of a trip comprises the processing costs of its serviced tasks and the traversal costs of the entities used for connecting these tasks.

The number of vehicles actually used is a decision variable in this problem. It is obvious that the VRP and the CARP are special cases of the NEARP.

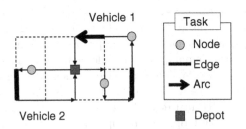

Fig. 1. NEARP in urban streets

2.2   A Precedent Solution for NEARP

Prins et al. solved the NEARP by a memetic algorithm (MA) which is composed of two phases. In the global phase, they use the selection and the crossover operations. In the local phase, they use a local search procedure instead of the mutation. Basically, the state of a solution is expressed as a set of trips corresponding to a fleet of vehicles and evaluated with total cost. When the solution is moved into the global phase, a set of trips is converted to a string, which is treated as a chromosome in MA. Note that required entities (nodes, edges, and arcs) are arranged according to the routing order but the delimiters partitioning trips are removed from the string when the conversion is carried out. In the global phase, two chromosomes are randomly selected first. The least cost chromosome is selected from these two as a parent in a population. Another parent is selected by the same manner. The extended order crossover is applied to the pair of parents in order to produce a pair of children. After one child is selected randomly from the pair, the string is divided into trips corresponding to individual vehicles by an optimal splitting procedure. Then, the trips are moved to the local phase. In the local phase, the local search procedure, which calls successive sub procedures listed as follows, is applied to the inside of a trip or between two trips. When a required edge is encoded as two opposite arcs, $inv(a)$ is the reversed task of a. The term 'task' is used as a required entity.

- Flip one task a, i.e., replace a by $inv(a)$ in its trip,
- Move one task a after another task or after the depot,
- Move two consecutive tasks a and b after another task or after the depot,
- Swap two tasks a and b,
- 2-opt moves

Moreover, each task a moved to another location or swapped with another task may be inserted as a or $inv(a)$. Each sub procedure ends by performing the first improving move detected or when all moves have been examined. The process is repeated until no further saving can be found. When the local optimization finishes, all trips are concatenated into a string in which delimiters are removed. After the resultant string is replaced by one of the previous chromosomes in the population, the global phase begins in the next generation.

3   The Proposed Method for Solving NEARP

Although the precedent method mentioned above shows good performance, its procedures are considerably complex. In particular, the local search procedure is complicated. It is reported that the local searches absorb 95% of the total MA running time on big instances. We propose a simpler data model and a one phase algorithm to solve the NEARP.

3.1   Data Modeling

The proposed solution modeling and algorithm are based on an internal network coding. In this coding, all entities (nodes, edges, arcs) are stored in a common form which is embodied as a three dimensional array. The first component of it expresses the head node of the entity and the second expresses the tail node. The third is the Boolean value that attains 1, if and only if the entity is an arc. If the head and the tail are same node, the entity is understood as a single node.

The model to express a solution of the NEARP is realized as a sequence of integers, i.e., a string. In the string, the position of a number, which is a symbol

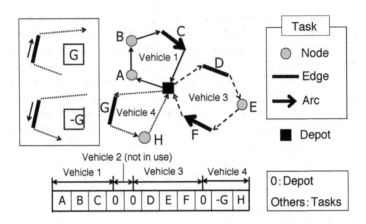

Fig. 2. The string model for NEARP

of the required entities (tasks), implies not only which vehicle tours the tasks but also the routing order of it. An example of the string model is illustrated in Fig.2. In the string, a required arc or a required node is expressed by a positive number, whereas a required edge is expressed by a positive number or a negative number according to the direction traversed. The special number '0' should be interpreted not only as the depot but also as the delimiter which partition the trips. If we denote the number of vehicles as m, $(m-1)$ '0' s are provided in the string. If there is no number between '0' and '0', the concerned vehicle is not in use.

When the cost of the solution is evaluated (see Sec.3.3), the internal coding corresponding to the task in the string is referred.

3.2   Transformation Rules

A new state of solution is generated from the present state by one of the following three types of transformation rules. The first rule is to exchange a number with another one in the string. The second rule is to delete an arbitrary number and then insert it to another position in the string. The third rule is to reverse the traversing direction of an undirected edge. These three transformation rules are illustrated in Fig.3. However, the third rule 'direction reversal' can not be applied to directed arcs.

Note that the rules are also applied to the special number '0'. In other words, '0' is evenly treated with other numbers. If 'one-to-one exchange' is executed within a substring, only a route of one vehicle will be changed.

If 'one-to-one exchange' is executed between two components that belong to different substrings, two tasks will be exchanged between two vehicles. An example is illustrated in Fig.4.

If 'one-to-one exchange' is executed between a non-zero number and '0', a partial route of one vehicle is moved and embedded in another vehicle's route. An example is illustrated in Fig.5.

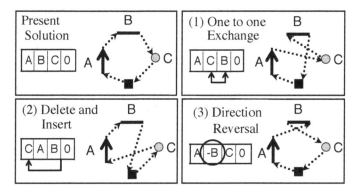

Fig. 3. Three transformation rules

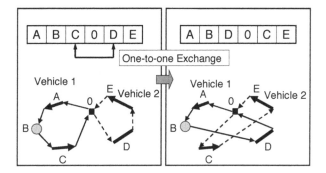

Fig. 4. A result of 'one-to-one exchange' between two non-zeros striding over '0'

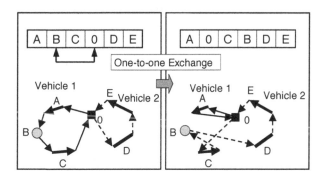

Fig. 5. A result of 'one-to-one exchange' between non-zero and '0'

When the second transformation rule 'delete and insert' is applied, several different cases also arise. If a non-zero number is deleted and inserted striding over '0', a task is moved to another vehicle's route. An example is illustrated in Fig.6.

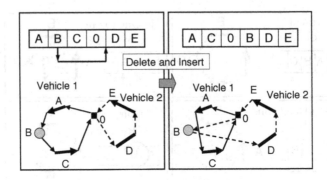

Fig. 6. A result of deleting non-zero and inserting it striding over '0'

If a '0' is deleted and inserted to another position, the routes are changed drastically. An example is illustrated in Fig.7.

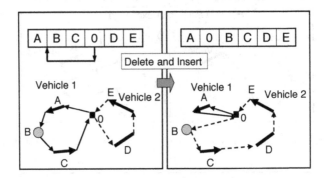

Fig. 7. A result of deleting '0' and inserting it to another position

3.3   Objective Function

The objective of the NEARP is the minimization of the total cost which is subject to constraints including the loading capacity of each vehicle. The objective function of the NEARP is formulated as follows.

$$E = \sum_{i=1}^{n} c_{s_i} + \sum_{i=1}^{n-1} p_{s_i, s_{i+1}} \qquad (1)$$

where $s = (s_1, s_2, \cdots, s_n)$ is a string that consists of the tasks and a depot; c_k is the cost for processing the task k (if $k = 0$, then $c_k = 0$); $p_{k,l}$ is the minimal traversing cost from the tail of the task k to the head of the task l.

In spite of the fact that the sum of processing costs is constant, the objective is defined as Eq.(1) in order to apply our method to the instances given by Prins et. al and compare our method with their method. All values of $p_{k,l}$ are calculated based on the internal network coding explained in Sec. 3.1 using the Warshall-Floyd's algorithm beforehand with optimization. If k and l are both the edge tasks, four different values of $p_{k,l}$ must be computed, according to the traversing directions of k and l.

3.4 Optimization Algorithm

Simulated annealing (SA) is adopted as the optimization technique since it is characterized by simple stochastic procedures and by global searching scope. Starting with a random initial state, it is expected to approach an equilibrium point. In the proposed method, we apply the three transformation rules described in Sec. 3.2 to the string model. The entire algorithm for the NEARP is described as follows.

{*I. Preparation*}
Read network data;
Calculate the minimum path cost $p_{k,l}$ between two tasks;
{*II. Initialization*}
Generate a random initial feasible solution x_0;
$x := x_0; \quad x^* := x;$
Calculate INITTEMP *by exploratory SA executions;*
$T :=$ INITTEMP*;*
{*III. Optimization by SA*}
Minimize E by applying the three transformation rules to the string model of x in the framework of SA;
{*IV. Output*}
Output the best solution x^*.

Step III, that is the main part of this algorithm, is detailed as follows.

Repeat
    trials := 0; *changes* := 0;
    Repeat
        trials := *trials* + 1;
        Generate a new state x' from the current state x by applying randomly one of the three transformation rules to the string model of x;
        If *x' is feasible* **Then**
            Calculate $\Delta E = E(x') - E(x)$;
            If $\Delta E < 0$ **Then**
                x' is accepted as a new state;
                If $E(x') < E(x^*)$ **Then** $x^* := x'$
            Else *x' is accepted with probability* $\exp(-\Delta E/T)$
            If *x' is accepted* **Then** *changes* := *changes* + 1; $x := x'$
    Until *trials* $\geq$ SIZEFACTOR $\cdot N$ *or changes* $\geq$ CUTOFF $\cdot N$;
    $T := T \cdot$ TEMPFACTOR
Until $T \leq$ INITTEMP/FIN_DIVISOR

The feasibility of the generated state x' is determined by checking whether the total load of any vehicle does not exceed its loading capacity. Parameters that appear in the above algorithm will be explained in Sec.4.3.

4 Experimental Evaluations

We have tried to solve the NEARP and compared it with the results of the precedent method.

4.1 Instances of NEARP

Prins et al. provide 23 instances of the NEARP [8]. These instances were produced by their original generator accompanied with randomization. They include 1-93 required nodes, 0-94 required edges and 0-149 required arcs, among 11-150 nodes and 71-311 links (integrated alias with edges and arcs). As mentioned by them, the lower bounds have not been found for their NEARP instances. Because the VRP and the CARP are special cases of the NEARP, they tried to test their solution method by applying it to the benchmark VRPs and CARPs. They obtained good results. The data files of NEARP instances were sent by them at our request.

4.2 Generation of a Random Initial Feasible Solution

The initial feasible solution affects the quality of the computational result. We generate a random initial feasible solution described as step II of our algorithm (Sec.3.4) by using the following procedures.

1. *Sort tasks in descending order of their quantity of demand;*
2. *Assign the sorted tasks to vehicles without exceeding their loading capacity;*
3. *Form the string model in accordance with the result of the assignment;*
4. *Apply random change according to one of the three transformation rules for 1000 times. Compute the feasibility rate, that is the rate of the number of the feasible solutions generated per the number of total generated solutions, of each transformation rule.*

4.3 Probabilities of Three Transformations and Parameter Setting

Three transformation rules mentioned in Sec.3.2 should be applied *equally probably* to produce a new feasible state of solution, according to the results of the preliminary numerical experiments.

However, the feasibility rates of the results generated from the three transformations are not equal. In order to produce a feasible solution generated by each transformation with almost equal probability, we adjust the rate of applying these transformations, taking account of the feasibility rates computed in the initialization procedures described above.

According to the preliminary experiments and the reference to the recommended values by Johnson et al. [11,12], the standard values of the parameters that appear in our simulated annealing algorithm are commonly set as follows.

$N = 2L^2$ (L : length of string)
$SIZEFACTOR = 4$
$CUTOFF = 0.1$ (Repeat iterations in the same T,
    until ($trials \geq SIZEFACTOR \cdot N$ or $changes \geq CUTOFF \cdot N$))
$INITPROB = 0.4$ (Initial acceptance probability)
$TEMPFACTOR = 0.99$ ($T_{n+1} = 0.99\,T_n$)
$FIN\_DIVISOR = 10$ (If $T \leq INITTEMP/FIN\_DIVISOR$,
    terminate the whole iterations.)

$INITTEMP$ must be calculated by exploratory and simplified SA executions in step II of the entire algorithm, so as to make $changes/trials = INITPROB$ ($=$ 0.4).

4.4   Experimental Results

A comparison between the solutions generated by the proposed method and the solutions given by Prins et al. is done for 23 NEARP instances. In Table 1, the column MA indicates the total cost computed with the memetic algorithm by Prins et al. $Best^*MA$ indicates the best solution found with various parameter settings during their experiments. Because SA is an optimization method based on randomization, we solved each problem ten times with the standard parameter setting shown above. $Avg_{10}SA$ is the averaged result of ten computations, while $Best_{10}SA$ is the best result of them. Note that the $Best_{10}SA$ is obtained by computations with the standard parameter setting and it is quite different from the $Best^*MA$ in spite of using the same word '$Best$'. Average deviations over the best values (indicated in bold-face) are given in percent.

Compared with the results of MA, $Avg_{10}SA$ obtains better solutions in 13 instances, and equal solutions in 2 instances. Compared with the results of MA, $Best_{10}SA$ obtains better solutions in 16 instances, and equal solutions in 4 instances. Compared with the $Best^*MA$, $Best_{10}SA$ obtains better results in 10 instances, and equal results in 5 instances in spite of using the common standard parameters. Regarding the average deviations over the best value, $Avg_{10}SA$ is superior to MA, $Best_{10}SA$ gets better results than $Best^*MA$.

We coded our program in the C language and computed on a 1.8GHz Pentium IV PC with Windows XP, while the program by Prins et al. was coded in the Pascal-like language Delphi version 5 and tested on a 1 GHz Pentium III PC with Windows 98. Despite taking the difference of computing speed into consideration, both the average and the deviation of the computation time spent for our method are smaller than for their method (Fig.8).

5   Varieties of NEARP

The proposed method is adaptable to additional conditions. In this section, we briefly introduce three possible varieties.

Table 1. Computational results for NEARP instances

No	Nodes	Tasks(N, E, A)	MA	MAtime(s)	$Best^*MA$	$Avg_{10}SA$	SAtime(s)	$Best_{10}SA$
1	21	(11,0,37)	2632	108.3	**2589**	2617.1	15.1	2595
2	68	(36,0,149)	12336	1078.5	12241	12322.4	661.4	**12220**
3	31	(16,8,55)	3702	157.0	3669	3695.2	56.0	**3660**
4	53	(10,75,13)	**7583**	548.1	**7583**	7728.5	76.1	7641
5	32	(23,4,23)	4562	100.0	4544	4685.3	41.5	**4531**
6	49	(40,4,64)	**7087**	204.5	**7087**	7101.4	98.0	**7087**
7	75	(54,8,106)	9974	662.6	9748	9704.8	351.7	**9615**
8	77	(63,6,108)	10714	767.6	10658	10710.2	263.8	**10524**
9	29	(6,39,5)	4041	140.8	**4038**	4132.4	12.5	4103
10	56	(4,94,9)	7755	843.2	**7582**	7763.2	108.3	7687
11	69	(65,6,11)	4503	414.7	4494	4599.6	49.8	4506
12	38	(1,0,52)	**3235**	71.3	**3235**	**3235.0**	21.4	**3235**
13	150	(79,2,60)	9339	550.6	**9110**	9270.6	312.8	9133
14	94	(93,0,0)	8615	357.2	**8566**	8769.3	65.3	8608
15	52	(0,91,0)	8359	390.2	8340	8385.3	97.3	**8280**
16	71	(36,0,133)	9389	536.1	8933	9024.3	445.5	**8886**
17	42	(16,16,31)	4165	116.1	**4037**	4107.6	43.0	**4037**
18	117	(39,0,88)	7411	475.7	7254	7214.6	278.4	**7098**
19	126	(61,9,142)	17036	1273.4	16554	16677.5	469.8	**16347**
20	43	(38,2,33)	4918	164.6	**4844**	4902.9	50.7	4846
21	60	(55,68,57)	18509	1370.6	18201	18318.3	530.4	**18069**
22	25	(7,10,25)	**1941**	65.8	**1941**	1970.5	9.5	**1941**
23	11	(3,2,15)	**780**	20.4	**780**	**780.0**	2.7	**780**
		Average	1.65%	452.9 (s)	0.38%	1.51%	176.6 (s)	0.22%

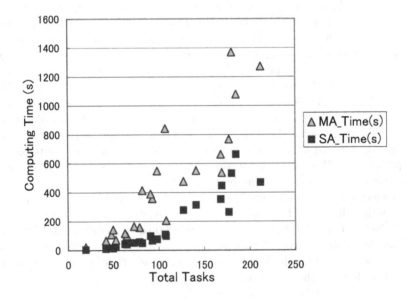

Fig. 8. Comparison of computation time

5.1   NEARP with Time Windows (NEARPTW)

If some of tasks are imposed with time windows, the NEARP is called as the NEARP with time windows (NEARPTW). To cope with the NEARPTW, we have incorporated a penalty term, concerning to sum of excess time over time window, into the cost function.

$$E' = aE + b\sum_{i=1}^{n} T_i \tag{2}$$

where E is total cost defined in Eq.(1), T_i is excess time over time window regarding the task i. Because the string model holds touring order of tasks and the arrival time of task i can be calculated, it is easy to compute T_i.

5.2   NEARP with Repeated Trips

We are able to deal with the case in which repetitive trips of a vehicle are allowed. After the first trip returns to the depot, load and unload are operated at the depot. Then, the second trip starts. In order to cope with this problem, we incorporate another delimiter, for example '999', into the string model. In the state transformation of a solution, '999' is treated the same as '0' as well as other numbers which represent tasks (Fig.9).

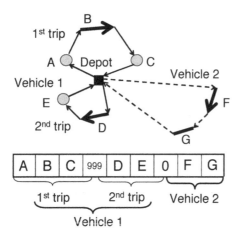

Fig. 9. NEARP with repeated trips

5.3   Multi Depots NEARP

The case in which more than one depot is managed at a time could be dealt with. We incorporate the other delimiter, for example '−999', into the string model. In the state transformation of a solution, '−999' is treated evenly with '0' and other numbers (Fig.10). In addition to these three examples, we are able to apply our method to the pickup and delivery problem and so on. These are implemented by incorporating special numbers into the string model.

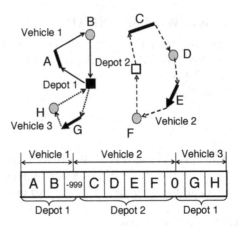

Fig. 10. Multi depots NEARP

6   Conclusion

The previous discussions are summarized as follows.

- The proposed method for solving the NEARP consists of a simple procedure based on the string data model. In the framework of simulated annealing, random applications of three transformation rules may occasionally cause the exchange of required tasks between two trips, the move of required tasks from one trip to another trip, and the exchange of tasks in the same trip and so on. Because transitions of a present state in the string data model are able to give drastic changes in solutions, fast convergence to an equilibrium point may be achieved.
- The solutions generated by the proposed method are compared with the solutions given by another method by making computational experiments with the NEARP instances. In most cases, the proposed method shows good performance.
- The proposed method is adaptable to additional conditions. Cases in which time windows are imposed, repetitive trips of a vehicle are allowed and more than one depot is managed at a time could all be dealt with.
- Although the proposed scheme is advantageous to complicated delivery operations, the following problems should be considered in order to apply the method to practical use.
  - The effects of 'one-to-one exchange' and 'delete and insert' in the whole string model should be analyzed theoretically.
  - Applications of the tabu search and the genetic algorithm to the proposed string model and the transformation rules should be attempted. The particular metaheuristics fit for the string model should be found out.
  - The proposed data model can be applied to NEARP with time windows, multiple depots cases and pickup and delivery cases. The application to these cases should be examined. However, the bench-mark problems and the necessary data of actual delivery cases have not been obtained yet.

References

1. Japan Petroleum Energy Center: Study on CO2 emissions from automobiles and refineries. In: CO2 emissions study WG report in 4th JCAP Conference (2005), http://www.pecj.or.jp/english/jcap/pdf/JCAP200506/2_3e.pdf
2. Kokubugata, H., Itoyama, H., Kawashima, H.: Vehicle Routing Methods for City Logistics Operations. In: Preprint for IFAC Symposium on Transportation Systems, pp. 727–732 (1997)
3. Hasama, T., Kokubugata, H., Kawashima, H.: A Heuristic Approach Based on the String Model to Solve Vehicle Routing Problem with Backhauls. In: Preprint for 5th Annual World Congress on Intelligent Transport Systems, No.3025 (1998)
4. Hasama, T., Kokubugata, H., Kawashima, H.: A Heuristic Approach Based on the String Model to Solve Vehicle Routing Problem with Various Conditions. In: Preprint for 6th World Congress on Intelligent Transport Systems, No.3027 (1999)
5. Hertz, A., Laporte, G., Mittaz, M.: A Tabu Search Heuristic for the Capacitated Arc Routing Problem. Operations Research 48, 129–135 (2000)
6. Lacomme, P., Prins, C., Ramdane-Cherif, W.: A Genetic Algorithm for the Capacitated Arc Routing Problem and its Extensions. In: Boers, E.J.W., Gottlieb, J., Lanzi, P.L., Smith, R.E., Cagnoni, S., Hart, E., Raidl, G.R., Tijink, H. (eds.) EvoIASP 2001, EvoWorkshops 2001, EvoFlight 2001, EvoSTIM 2001, EvoCOP 2001, and EvoLearn 2001. LNCS, vol. 2037, pp. 473–483. Springer, Heidelberg (2001)
7. Kokubugata, H., Hirashima, K., Kawashima, H.: A Practical Solution of Capacitated Arc Routing for City Logistics. In: Proceeding of 11th IFAC Symposium on Control in Transportation Systems, No.222 (2006)
8. Prins, C., Bouchenoua, S.: A Memetic Algorithm Solving the VRP, the CARP and more General Routing Problems with Nodes, Edges and Arcs. In: Hart, W., Kranogor, N., Smith, J. (eds.) Recent Advances in Memetic Algorithms, Studies in Fuzziness and Soft Computing, vol. 166, pp. 65–85. Springer, Heidelberg (2004)
9. Oppen, J., Lokketangen, A.: Arc Routing in a Node Routing Environment. Computers & Operations Research 33, 1033–1055 (2006)
10. Crescenzi, P., Kann, V.: A Compendium of NP Optimization Problem. Web site: http://www.nada.kth.se/ viggo/wwwcompendium/node103.html
11. Johnson, D.S., Aragon, C.R., McGeoch, L.A., Schevon, C.: Optimization by Simulated Annealing: An Experimental Evaluation, part I, graph partitioning. Operations Research 37, 865–892 (1989)
12. Johnson, D.S., Aragon, C.R., McGeoch, L.A., Schevon, C.: Optimization by Simulated Annealing: An Experimental Evaluation, part II, graph coloring and number partitioning. Operations Research 39, 378–406 (1991)

Probabilistic Beam Search for the Longest Common Subsequence Problem[*]

Christian Blum and Maria J. Blesa

ALBCOM, Dept. Llenguatges i Sistemes Informàtics,
Universitat Politècnica de Catalunya, Barcelona, Spain
{cblum,mjblesa}@lsi.upc.edu

Abstract. Finding the common part of a set of strings has many important applications, for example, in pattern recognition or computational biology. In computer science, this problem is known as the longest common subsequence problem. In this work we present a probabilistic beam search approach to solve this classical problem. To our knowledge, this algorithm is the first stochastic local search algorithm proposed for this problem. The results show the great potential of our algorithm when compared to existing heuristic methods.

1 Introduction

The longest common subsequence (LCS) problem is one of the classical string problems. Given a problem instance $(\mathcal{S}, \Sigma)$, where $\mathcal{S} = \{s_1, s_2, \ldots, s_n\}$ is a set of n strings over a finite alphabet Σ, the problem consists in finding a longest string t^* that is a subsequence of all the strings in $\mathcal{S}$. Such a string t^* is called a *longest common subsequence* of the strings in $\mathcal{S}$. Note that a string t is called a subsequence of a string s, if t can be produced from s by deleting characters. For example, *dga* is a subsequence of *adagtta*. The LCS problem is in general NP-hard [1]. In case $n = 2$ the problem is polynomially solvable, for example, by dynamic programming [2]. Traditional applications of this problem are in data compression, syntactic patter recognition, and file comparison [3], whereas more recent applications also include computational biology [4].

To our knowledge, stochastic local search methods have not been applied so far to the LCS problem. Existing approximate methods fall into the class of deterministic heuristics. Examples are simple heuristics based on sequential solution construction (numerous references can be found in [5]), or more sophisticated methods such as the so-called expansion algorithm [6]. The relative performance of these heuristics depends on the characteristics of the problem instances. In this work we propose a so-called probabilistic beam search (Pbs) approach to tackle the LCS problem. Pbs is a stochastic local search algorithm based on solution construction and lower bounding techniques. It is a probabilistic version of

[*] This work was supported by grant TIN-2005-08818-C04-01 (OPLINK) of the Spanish government, and by the *Ramón y Cajal* program of the Spanish Ministry of Science and Technology of which Christian Blum is a research fellow.

T. Stützle, M. Birattari, and H.H. Hoos (Eds.): SLS 2007, LNCS 4638, pp. 150–161, 2007.
© Springer-Verlag Berlin Heidelberg 2007

deterministic beam search, which is an incomplete derivative of branch & bound. In [7] we proposed a similar technique for a related problem, called the shortest common supersequence problem (SCSP). The success of the application to the SCSP was the motivation for applying this approach also to the LCS problem.

This paper is organized as follows. In section 2 we present the constructive heuristic being the basis of our PBS algorithm. In section 3 we present the PBS algorithm, and in section 4 we describe the experimental evaluation of our approach. Finally, in section 5 we offer conclusions and an outlook to future work.

2   A Sequential Construction Heuristic

In this section we outline the so-called BEST-NEXT heuristic [8,9], which is a fast heuristic for the LCS problem. We will use the construction mechanism of this heuristic for our probabilistic beam search approach. Given a problem instance $(\mathcal{S}, \Sigma)$, the BEST-NEXT heuristic produces a common subsequence t sequentially from left to right by appending at each construction step a letter to t such that t maintains the property of being a common subsequence of all strings in $\mathcal{S}$. The algorithm is pseudo-coded in Algorithm 1.. In order to explain all the components of the algorithm we need to introduce the following definitions and notations. They all assume a given common subsequence t of the strings in $\mathcal{S}$.

1. Let $s_i = s_i^A \cdot s_i^B$ be the partition of s_i into substrings s_i^A and s_i^B such that t is a subsequence of s_i^A, and s_i^B has maximal length. Given this partition, which is well-defined, we introduce position pointers $p_i := |s_i^A|$ for $i = 1, \dots, n$ (see Figure 1 for an example).
2. The position of the first appearance of a letter $a \in \Sigma$ in a string $s_i \in \mathcal{S}$ after the position pointer p_i is well-defined and denoted by p_i^a. In case a letter $a \in \Sigma$ does not appear in s_i^B, p_i^a is set to ∞ (see Figure 1).
3. A letter $a \in \Sigma$ is called *dominated*, if exists at least one letter $b \in \Sigma$, $a \neq b$, such that $p_i^b < p_i^a$ for $i = 1, \dots, n$;
4. $\Sigma_t^{\mathrm{nd}} \subseteq \Sigma$ henceforth denotes the set of non-dominated letters of the alphabet Σ. Moreover, for all $a \in \Sigma_t^{\mathrm{nd}}$ it is required that $p_i^a < \infty$. Hence, we require that in each string s_i a letter $a \in \Sigma_t^{\mathrm{nd}}$ appears at least once after position pointer p_i.

Function ChooseFrom(Σ_t^{nd}) is used to choose at each iteration one of the letters from Σ_t^{nd} for extending the common subsequence t. This is done by means of a so-called *greedy function*. In the following we present two different greedy functions that may be used. The first one—henceforth denoted by $\eta_1(\cdot)$—is known from the literature (see, for example, [8]). The second one—henceforth denoted by $\eta_2(\cdot)$—is a new development of this work. They are defined as follows:

$$\eta_1(a) = \min\{|s_i| - p_i^a \mid i = 1, \dots, n\} \quad, \forall\, a \in \Sigma_t^{\mathrm{nd}} \tag{1}$$

$$\eta_2(a) = \left(\sum_{i=1,\dots,n} \frac{p_i^a - p_i}{|s_i^B|} \right)^{-1} \quad, \forall\, a \in \Sigma_t^{\mathrm{nd}} \tag{2}$$

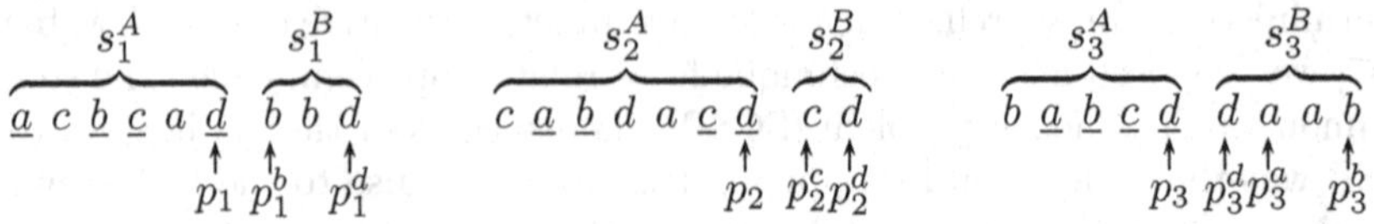

(a) String s_1 (b) String s_2 (c) String s_3

Fig. 1. Given is the problem instance $(S = \{s_1, s_2, s_3\}, \Sigma = \{a, b, c, d\})$ where $s_1 = acbcadbbd$, $s_2 = cabdacdcd$, and $s_3 = babcddaab$. Let us assume that $t = abcd$. (a), (b), and (c) show the corresponding division of s_i into s_i^A and s_i^B, as well as the setting of the pointers p_i and the next positions of the 4 letters in s_i^B. Note that in case a letter does not appear in s_i^B, the corresponding pointer is set to ∞. For example, as letter a does not appear in s_1^B, we set $p_1^a := \infty$.

Algorithm 1. BEST-NEXT heuristic for the LCS problem

1: **input:** a problem instance (S, Σ)
2: **initialization:** $t := \epsilon$ (where ϵ is the empty string)
3: **while** $|\Sigma_t^{nd}| > 0$ **do**
4:     $a := \mathsf{ChooseFrom}(\Sigma_t^{nd})$
5:     $t := ta$
6: **end while**
7: **output:** common subsequence t

In function $\mathsf{ChooseFrom}(\Sigma_t^{nd})$ we choose $a \in \Sigma_t^{nd}$ such that $\eta_j(a) \geq \eta_j(b)$ for all $b \in \Sigma_t^{nd}$.[1] This completes the description of the BEST-NEXT heuristic.

3   Probabilistic Beam Search for the LCS Problem

In the following we present a probabilistic beam search (PBS) approach for the LCS problem. Our algorithm is based on the construction mechanism of the BEST-NEXT heuristic outlined in the previous section. Beam search is a classical tree search method that was introduced in the context of scheduling [10]. The central idea behind beam search is to allow the extension of partial solutions (here: subsequences) in more than one way. The version of beam search that we implemented—see Algorithm 2.—works as follows: The algorithm requires, apart from a problem instance (S, Σ), three input parameters. $k_{bw} \in \mathbb{Z}^+$ is the so-called *beam width*, $\mu \in \mathbb{R}^+ \geq 1$ is a parameter that is used to determine the number of children that can be chosen at each step, and t_{bsf} is the best solution found so far. At each step of the algorithm is given a set B of subsequences which is called the *beam*. At the start of the algorithm B only contains the empty string ϵ (that is, $B := \{\epsilon\}$). Let C denote the set of all possible extensions

---

[1] Note that for each application of the BEST-NEXT heuristic, we either use $j = 1$ or $j = 2$.

Algorithm 2. Probabilistic beam search (PBS) for the LCS problem

1: **input:** a problem instance $(\mathcal{S}, \Sigma)$, k_{bw}, μ, t_{bsf}
2: $B_{compl} := \{\epsilon\}$, $B := \{\epsilon\}$
3: **while** $B \neq \emptyset$ **do**
4:    $C := \mathsf{Children\_of}(B)$
5:    $B := \emptyset$
6:    **for** $k = 1, \ldots, \min\{\lfloor \mu \cdot k_{bw} \rfloor, |C|\}$ **do**
7:       $\langle t, a \rangle := \mathsf{Choose\_From}(C)$
8:       $t := ta$
9:       **if** $\mathrm{UB}(t) = |t|$ **then**
10:          $B_{compl} := B_{compl} \cup \{t\}$
11:          **if** $|t| > |t_{bsf}|$ **then** $t_{bsf} := t$ **end if**
12:       **else**
13:          **if** $\mathrm{UB}(t) \geq |t_{bsf}|$ **then** $B := B \cup \{t\}$ **end if**
14:       **end if**
15:       $C := C \setminus \{t\}$
16:    **end for**
17:    $B := \mathsf{Reduce}(B, k_{bw})$
18: **end while**
19: **output:** $\mathrm{argmax}\{|t| \mid t \in B_{compl}\}$

(children) of the subsequences in B.[2] At each step, $\lfloor \mu \cdot k_{bw} \rfloor$ different extensions from C are selected; each selection step is either performed probabilistically, or deterministically. A chosen extension, which is also a subsequence, is either stored in set B_{compl} in case it is a complete solution, or—in case its upper bound value $\mathrm{UB}(\cdot)$ is greater than the length of the best-so-far solution t_{bsf}—it is stored in the new beam B. At the end of each step the new beam B is reduced in case it contains more than k_{bw} (the *beam width*) subsequences. This is done by evaluating the subsequences in B by means of the upper bound $\mathrm{UB}(\cdot)$ mentioned already above, and by subsequently selecting the k_{bw} subsequences with the greatest upper bound values.

In the following we explain some aspects of Algorithm 2. in more detail. The algorithm uses three different functions. Given the current beam B as input, function $\mathsf{Children\_of}(B)$ produces the children (extensions) of all the subsequences in B in the form of a set C of tuples $\langle t, a \rangle$ for each pair $t \in B$ and $a \in \Sigma_t^{nd}$.

The second function—$\mathsf{Choose\_From}(C)$—is used for choosing one of the (remaining) children in C. This is done by means of one of the greedy functions outlined in the previous section. However, note that the weights given by a greedy function are only meaningful for the comparison of two children $\langle t, a \rangle$ and $\langle t, b \rangle$, while they are meaningless for the comparison of two children $\langle t, a \rangle$ and $\langle z, b \rangle$ (where $t \neq z$). We solved this problem as follows. First, instead of the weights given by a greedy function, we use the corresponding ranks. More

---

[2] Remeber that the construction mechanism of the BEST-NEXT heuristic is based on extending a subsequence t by appending one letter from Σ_t^{nd}.

in detail, given all children $\{\langle t, a\rangle \mid a \in \Sigma_t^{\mathrm{nd}}\}$ descending from a subsequence t, the child $\langle t, b\rangle$ with $\eta_j(\langle t, b\rangle) \geq \eta_j(\langle t, a\rangle)$ for all $a \in \Sigma_t^{\mathrm{nd}}$ receives rank 1, denoted by $r_j(\langle t, b\rangle) = 1$.[3] The child with the second highest greedy weight receives rank 2, and so on. Note that here we extend the notation $\eta_j(a)$ (as introduced in the previous section) to $\eta_j(\langle t, a\rangle)$, with the same meaning. Next, for evaluating a child $\langle t, a\rangle$ we use the sum of the ranks of the greedy weights that correspond to the construction steps performed to construct subsequence ta, that is

$$\nu(\langle t, a\rangle) := r_j(\langle \epsilon, t_1\rangle) + \left(\sum_{i=1}^{|t|-1} r_j(\langle t_1...t_i, t_{i+1}\rangle) \right) + r_j(\langle t, a\rangle) \quad , \tag{3}$$

where ϵ is the empty string, and t_i denotes the letter on position i in subsequence t. Morever, $t_1...t_i$ denotes the substring of t from position 1 to postion i. In contrast to the greedy function weights, these newly defined $\nu(\cdot)$-values can be used to compare children descending from different subsequences, which is done as follows. Given the set of children C, each $\langle t, a\rangle \in C$ is assigned the following probability:

$$\mathbf{p}(\langle t, a\rangle) := \frac{\nu(\langle t, a\rangle)}{\sum_{\langle z, b\rangle \in C} \nu(\langle z, b\rangle)} \quad , \forall \langle t, a\rangle \in C \tag{4}$$

When function Choose_From(C) is called, it is first decided whether to perform the choice of a child determinstically, or probabilistically. This is done by drawing a random number $q \in [0, 1]$ and comparing it to a parameter $0 \leq d \leq 1$. If $q < d$, the child $\langle t, a\rangle$ to be returned by function Choose_From(C) is selected such that $\langle t, a\rangle := \mathrm{argmax}\{\mathbf{p}(\langle z, b\rangle) \mid \langle z, b\rangle \in C\}$. Otherwise, the child to be returned is selected by roulette-wheel-selection according to the probabilities defined in Equation 4.

The last function used by the PBS algorithm is Reduce(B, k_{bw}). In case $|B| > k_{\mathrm{bw}}$, this function removes from B step-by-step those subsequences t that have an upper bound value UB(t) smaller or equal to the upper bound value of all the other subsequences in B. The removal process stops once $|B| = k_{\mathrm{bw}}$. Remember that k_{bw} is the beam width of the PBS algorithm, that is, the maximum number of subsequences that the algorithm is allowed to consider for extension.

Finally, we outline the upper bound function UB$(\cdot)$ that we used in our algorithm. Remember that a given subsequence t splits each string $s_i \in \mathcal{S}$ into a first part s_i^A and into a second part s_i^B, that is, $s_i = s_i^A \cdot s_i^B$ (see previous section). Henceforth, $|s_i^B|_a$ denotes the number of occurences of letter a in s_i^B for all $a \in \Sigma$. Then,

$$\mathrm{UB}(t) := |t| + \sum_{a \in \Sigma} \min\{|s_i^B|_a \mid i = 1, \ldots, n\} \quad . \tag{5}$$

In words, for each letter $a \in \Sigma$ we take the minimum of the occurences of a in s_i^B, $i = 1, \ldots, n$. Summing up these minima and adding the result to the length of t results in the upper bound. This completes the description of the PBS algorithm.

---

[3] Remember that $j \in \{1, 2\}$ indicates the used greedy function.

3.1 Deterministic Versus Probabilistic Beam Search

In this work we study two different ways of using the PBS algorithm: (1) Deterministic beam search, and (2) probabilistic beam search used in a multi-start fashion.

Deterministic beam search—henceforth simply denoted by BS—is obtained from algorithm PBS by setting $d = 1$. With this setting each choice of a child from C is done deterministically. Moreover, the best-so-far solution that is required as input of PBS is set to the empty string ϵ, that is, $t_{bsf} := \epsilon$.

A more interesting use of PBS is achieved by a setting of $d < 1$, and by using PBS in a multi-start fashion (see Algorithm 3.). The resulting algorithm—see Algorithm 3.—is denoted by MS-PBS. As stopping condition we use a maximum number of iterations, that is, calls of Algorithm 2.

Algorithm 3. Multi-start probabilistic beam search (MS-PBS)

1: **input:** a problem instance $(\mathcal{S}, \Sigma)$, k_{bw}, μ
2: $t_{bsf} := \epsilon$
3: **while** stopping conditions not satisfied **do**
4:    $t := \text{PBS}((\mathcal{S}, \Sigma), k_{bw}, \mu, t_{bsf})$ {see Algorithm 2.}
5:    **if** $|t| > |t_{bsf}|$ **then** $t_{bsf} := t$
6: **end while**
7: **output:** t_{bsf}

---

4 Experimental Evaluation

Apart from MS-PBS and BS, we also included the following algorithms in the experimental evaluation. The BEST-NEXT heuristic was applied with greedy function $\eta_1(\cdot)$ (see Equation 1), as well as with greedy function $\eta_2(\cdot)$ (see Equation 2). The former version is henceforth denoted by BN(1) and the latter one by BN(2). Morever, we applied to all instances the expansion algorithm [6], henceforth denoted by EXP. We implemented all these algorithms in ANSI C++ using GCC 3.2.2 for compiling the software. Our experimental results were obtained on a PC with an AMD64X2 4400 processor and 4 Gb of memory.

4.1 Parameter Setting of Probabilistic Beam Search

Remember that Algorithm 2. has 4 paramters: k_{bw} is the beam width, that is, the maximum number of subsequences that the algorithm is allowed to keep at any time; μ is used to determine the number of children that can be chosen from set C at each step of the algorithm; d is the probability that determines the balance between probabilistic and deterministic choices of children from C;[4] finally, we have to decide between greedy functions $\eta_1(\cdot)$ and $\eta_2(\cdot)$.

---

[4] When $d = 0$ the algorithm is completely random, while a setting of $d = 1$ results in deterministic beam search.

For the experimental evaluation presented in this paper we decided to fix the parameters after an initial tuning by hand. A thorough tuning process is left for future work. We decided for a setting of $k_{bw} = 10$ and $\mu = 3$.[5] Concerning the choice of the greedy function, we decided for $\eta_1(\cdot)$ for the application to the instances of Set1, whereas we chose $\eta_2(\cdot)$ for the instances of Set2 (see Section 4.2 for a description of the problem instances). As we will explain in the following section, $\eta_1(\cdot)$ seems to work much better than $\eta_2(\cdot)$ for instances from Set1, whereas the opposite is the case for instances of Set2. Finally, after tuning by hand we chose a setting of $d = 0.8$, which means that on average 20% of the choices of children from set C are performed probabilistically.

4.2   Benchmark Instances

Two different sets of benchmark instances have been used in the experimentation. The instances of the first set—henceforth denoted by Set1—were produced such that the length of an optimal solution is known. All instances of Set1 have the following characteristics: Each of them consists of 10 strings of length $l \in \{100, 200, \ldots, 1000\}$, derived from an alphabet Σ, where $|\Sigma| \in \{2, 4, 8, 24\}$. For each combination of l and $|\Sigma|$ we produced randomly 10 instances as follows. The first 2 strings of each instance are randomly generated. Then, dynamic programming is applied in order to produce the longest common subsequence t^* of these 2 strings. The 8 remaining strings of an instance are produced on the basis of t^*, which is extended by adding randomly chosen letters from Σ at random positions until length l is reached. As there are 40 combinations of l and $|\Sigma|$, we produced a total of 400 instances. Note that due to the above outlined construction mechanism, the length of an optimal solution of an instance I is $|t^*|$.

The second set of instances—henceforth denoted by Set2—was generated quite differently. Given $h \in \{100, 200, \ldots, 1000\}$ and Σ (where $|\Sigma| \in \{2, 4, 8, 24\}$), an instance is produced as follows. First, a string s of length h is produced randomly from the alphabet Σ. The string s is in the following called *base string*. Each instance contains again 10 strings. Each of these strings is produced from the base string s by traversing s and by deciding for each letter with a probabilitiy of 0.1 whether to remove it, or not. Note that the 10 strings of such an instance are not necessarily of the same length. As we produced 10 instances for each combination of h and $|\Sigma|$, 400 instances were generated in total. Note that the values of optimal solutions of these instances are unknown. However, a lower bound is obtained as follows. While producing the 10 strings of an instance, we record for each position of the base string s, whether the letter at that position was removed for the generation of at least one of the 10 strings. The number of positions in s that were never removed constitutes the lower bound value henceforth denoted by LB_I with respect to an instance I.

---

[5] The only exception occurs when instances with $|\Sigma| = 2$ are concerned. In this case we chose $\mu = 1.5$, because a setting of $\mu = 2$ would mean that at each step all children of C are selected.

4.3   Results for **Set1**

We applied each of the 5 algorithms exactly once to each problem instance. The Ms-Pbs algorithm was applied with 100 applications of algorithm Pbs. We present the results averaged over the 10 instances for each combination of l (the length of the 10 strings per instance), and the alphabet size $|\Sigma|$. Two measures are presented:

1. The (average) length of the solutions, expressed in percent with respect to the length of the optimal solutions. That is, 100% performance is achieved when the length of a solution is equal to the length of an optimal solution.
2. The computation time of the algorithms. In case of Ms-Pbs, this refers to the time the best solution was found within the 100 applications of algorithm Pbs (averaged over the 10 instances of the same type).

The results are shown graphically in Figure 2. The graphics on the left hand side show the algorithm performance (first measure), and the graphics on the right hand side show the computation times. The results allow us to draw the following conclusions. First, the performance of Bn(1) and Exp strongly decreases with increasing alphabet size. Surprisingly, the performance of the other three algorithms increases with increasing alphabet size. Second, algorithm Ms-Pbs is clearly the winner of the comparison. When $|\Sigma| > 4$, the algorithm nearly always finds the optimal solutions, regardless of l. Moreover, when the alphabet size is rather small and l is large the algorithm is clearly better than the 4 competitors. In particular, Ms-Pbs is never worse than Bs, its deterministic version. With respect to computation times, algorithm Exp clearly uses much more computation time than the other approaches, while the heuristics are much faster than the other approaches. Algorithms Bs and Ms-Pbs produce their results generally within a few seconds.

4.4   Results for **Set2**

We applied each of the 5 algorithms exactly once to each problem instance. The Ms-Pbs algorithm was again applied with 100 applications of algorithm Pbs. We present the results averaged over the 10 instances for each combination of h (the length of the base string that was used to produce an instance), and the alphabet size $|\Sigma|$. Apart from the average computation time (see previous section) we present the following measure: The (average) length of the solutions expressed in percent deviation from the respective lower bounds, which is computed as follows:

$$\left(\frac{f}{\text{LB}_I} - 1 \right) \cdot 100 \ , \tag{6}$$

where f is the length of the solution achieved by the respective algorithm. The results are shown graphically in Figure 3. The graphics on the left hand side show again the algorithm performance (in percentage deviation from the lower bound), and the graphics on the right hand side show the computation times. The following observations are of interest. First, in contrast to the results for

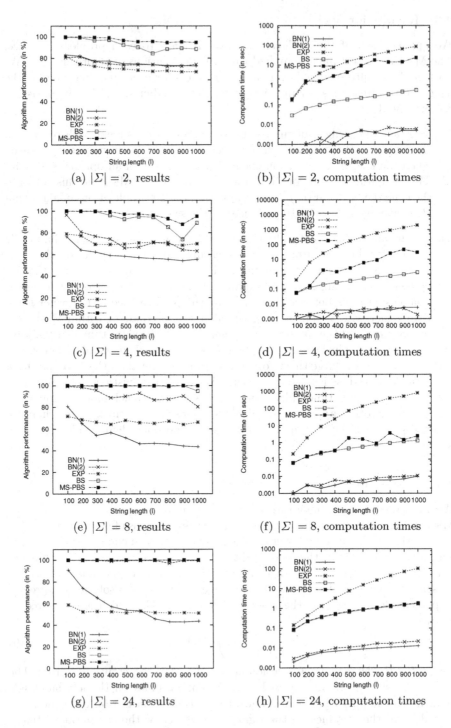

(a) $|\Sigma| = 2$, results

(b) $|\Sigma| = 2$, computation times

(c) $|\Sigma| = 4$, results

(d) $|\Sigma| = 4$, computation times

(e) $|\Sigma| = 8$, results

(f) $|\Sigma| = 8$, computation times

(g) $|\Sigma| = 24$, results

(h) $|\Sigma| = 24$, computation times

Fig. 2. Results for the instances of Set1

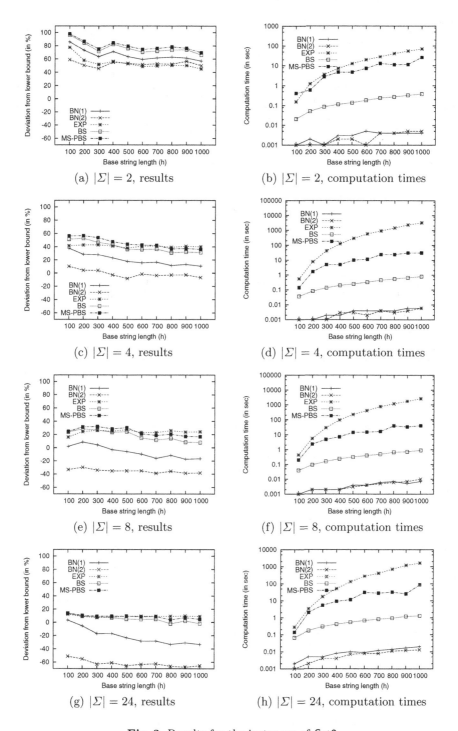

Fig. 3. Results for the instances of Set2

instance set Set1, algorithms BN(1) and EXP perform (in comparison) much better on the instances of Set2. Algorithm EXP is even slightly better than MS-PBS when $|\Sigma| \geq 4$ and h is large. However, EXP uses much more computation time than MS-PBS. Second, heuristic BN(1) is clearly better than heuristic BN(2) on this instance set. Concerning the beam search approaches, MS-PBS is again consistently better than BS, which shows the potential of a probabilistic beam search approach. However, especially when the alphabet size grows and h is large, there is room for improvement, which is shown by the fact that EXP is slightly better than MS-PBS in these cases. This might also be an indication that MS-PBS is not proporly tuned for these combinations of $|\Sigma|$ and h.

5   Conclusions and Future Work

In this work we presented a probabilistic beam search approach for the longest common subsequence problem, which is one of the classical string problems with applications in pattern recognition and computational biology. We generated two sets of each 400 problem instances. In summary, the results indicated that the probabilistic beam search algorithm is clearly the best algorithm for the first set of instances, whereas it is—with the current parameter settings—slightly inferior by the expansion algorithm on certain instances from the second set. Concerning computation times, the probabilistic beam search approach needs much less computation time for reaching—in most cases improving—the quality level of the expansion algorithm.

Future work will focus on two lines. First, we intend to improve probabilistic beam search in particular for the cases in which it performed slightly worse than the expansion algorithm. Possible directions include the implementation of a lookahead mechanism and the addition of a learning component, as for example in Beam-ACO [11]. The second line of future work concerns the problem instances. We plan to test our algorithms also for instances with a varying string number. In this study we only regarded instances with 10 strings. Moreover, we would like to learn the reasons for the different behaviour of the greedy functions and the expansion algorithm when comparing between the two sets of differently generated problem instances.

References

1. Maier, D.: The complexity of some problems on subsequences and supersequences. Journal of the ACM 25, 322–336 (1978)
2. Gusfield, D.: Algorithms on Strings, Trees, and Sequences. Computer Science and Computational Biology. Cambridge University Press, Cambridge (1997)
3. Aho, A., Hopcroft, J., Ullman, J.: Data structures and algorithms. Addison-Wesley, Reading, MA (1983)
4. Smith, T., Waterman, M.: Identification of common molecular subsequences. Journal of Molecular Biology 147(1), 195–197 (1981)

5. Bergeroth, L., Hakonen, H., Raita, T.: A survey of longest common subsequence algorithms. In: Proceedings of SPIRE 2000 – 7th International Symposium on String Processing and Information Retrieval, pp. 39–48. IEEE press, Los Alamitos (2000)
6. Bonizzoni, P., Della Vedova, G., Mauri, G.: Experimenting an approximation algorithm for the LCS. Discrete Applied Mathematics 110, 13–24 (2001)
7. Blum, C., Cotta, C., Fernández, A.J., Gallardo, J.E.: A probabilistic beam search algorithm for the shortest common supersequence problem. In: C.C., et al. (eds.) Proceedings of EvoCOP 2007 – Seventh European Conference on Evolutionary Computation in Combinatorial Optimisation. LNCS, vol. 4446, Springer, Heidelberg (2007) (In press)
8. Fraser, C.B.: Subsequences and supersequences of strings. PhD thesis, University of Glasgow (1995)
9. Huang, K., Yang, C., Tseng, K.: Fast algorithms for finding the common subsequences of multiple sequences. In: Proceedings of the International Computer Symposium, pp. 1006–1011. IEEE press, Los Alamitos (2004)
10. Ow, P.S., Morton, T.E.: Filtered beam search in scheduling. International Journal of Production Research 26, 297–307 (1988)
11. Blum, C.: Beam-ACO—Hybridizing ant colony optimization with beam search: An application to open shop scheduling. Computers & Operations Research 32(6), 1565–1591 (2005)

A Bidirectional Greedy Heuristic for the Subspace Selection Problem*

Dag Haugland

Department of Informatics, University of Bergen, Bergen, Norway
dag.haugland@ii.uib.no

Abstract. The Subspace Selection Problem (SSP) amounts to selecting t out of n given vectors of dimension m, such that they span a subspace in which a given target $b \in \Re^m$ has a closest possible approximation. This model has numerous applications in e.g. signal compression and statistical regression. It is well known that the problem is NP-hard. Based on elements from a forward and a backward greedy method, we develop a randomized search heuristic, which in some sense resembles variable neighborhood search, for SSP. Through numerical experiments we demonstrate that this approach has good promise, as it produces good results at modest computational cost.

Keywords: Subspace selection, dimension reduction, greedy methods, computational experiments.

1 Introduction

The *Subspace Selection Problem* (SSP) is a sparse approximation problem with applications in statistical regression, signal compression and other fields. It addresses the challenge of representing a high-dimensional target vector in low-dimensional space while minimizing the loss of information. As it is probably best understood in a signal compression context, we use this setting to introduce the problem.

Representing a sampled digital signal of dimension m in terms of two arrays of significantly smaller length, possibly with some information loss, can be accomplished by means of a large collection of elementary signal vectors of dimension m. By selecting t of the vectors, where $t < m/2$, and assigning a weight to each of them, a compressed version of the sampled signal is represented by the vector indices and the weights. When transmitted over a communication channel, transmission time is hence reduced if the t index and weight pairs are sent. Given that the collection of elementary signals are available at the reception side as well, the transmitted information can be used to restore an approximation to the sampled signal. The quality of the restored signal, however, depends largely on how the selection was made.

In this work, the elementary vectors are referred to as *atoms*, and a matrix in which the atoms are the columns is referred to as a *dictionary*. For given integers

---

* Supported by The Research Council of Norway under contract 160233/V30.

T. Stützle, M. Birattari, and H.H. Hoos (Eds.): SLS 2007, LNCS 4638, pp. 162–176, 2007.

m, n and $t \leq \min\{m, n\}$, target vector $b \in \Re^m$, and dictionary $A \in \Re^{m \times n}$, the problem is to find a linear combination of (at most) t columns of A such that its distance from b is minimized. An overview of applications of variants of this problem, along with theoretical results and algorithms, can be found in the monograph of Miller [1]. In signal compression applications, dictionaries are often *overcomplete*, which means that they have rank m and more than m columns. This assumption is however not general, applications to statistical regression [1] and classification (Couvreur and Bresler [2]) include cases where $n < m$.

Dimension reduction based on dictionaries is somehow related to traditional signal compression techniques based on orthogonal transformations of the target vector (transform coding). If $m = n$, A is orthogonal, and most components of $x = A^T b$ are small in absolute value, an approximation to b is found as $A\tilde{x}$. Here $\tilde{x}$ equals x in the t components with largest absolute value, whereas it is zero in all other components. Hence the target b is represented by the sparse vector $\tilde{x}$, implying an information loss depending on the choice of A. In *Principal Component Analysis*, which can be seen as a variant of transform coding, the transformation A is not given as an external parameter, but is rather based on the singular value decomposition of a matrix whose columns are the targets to be compressed.

The subspace selection problem is distinguished from transform coding in that the optimal solution is not identified solely by the magnitude of the atom coefficients. Transform coding can hence be understood as a simple greedy approach to the subspace selection problem, where the ordering of the atoms is fixed.

Owing to the computational intractability of the problem, the subspace selection problem has traditionally been attacked by heuristic methods aiming for solutions that are nearly optimal. Noteworthy among these are the variants of forward greedy algorithms by Mallat and Zhang [3], Pati et al. [4], Gharavi-Alkhansari and Huang [5], and the method of Natarajan [6]. All these approaches consider the distance to be given by the Euclidean measure, whereas the method by Haugland and Storøy [7] is a greedy algorithm based on a unit norm distance. A common feature of the forward greedy approaches is that a current subset of atoms, initialized to the empty set, is gradually extended until its cardinality becomes t. Since the set extension is non-trivial, the methods differ in the way this step is carried out. Opposed to the forward methods is the backward greedy approach suggested in [2]. This method, which is valid if the dictionary is not overcomplete, starts with accepting all $n > t$ atoms, and removes one in each iteration until only t remains.

In this paper, the simple ideas of forward and backward greedy algorithms are combined. We develop a bidirectional greedy heuristic that systematically alternates between moves in its primary neighborhood, which correspond to extensions by one, and moves in the secondary neighborhood (reductions by one). Likewise, we suggest a method where the roles of the neighborhoods are interchanged. This leads to two variants of a bidirectional greedy heuristic, distinguished by their primary search direction. An extension of the idea, where the approval of secondary moves is randomized, is also presented.

The paper is organized as follows: In the next section, we introduce some notation and background material, and define the problem in rigorous terms. In Sect. 3, we study how to extend or reduce a subset by one atom. Our heuristics utilizing such moves are developed in Sect. 4, and numerical experiments are given in Sect. 5.

2   Preliminaries

Let the columns of the dictionary $A \in \Re^{m \times n}$ be denoted $a^1, \ldots, a^n \in \Re^m$, and assume that any set of m columns are linearly independent. For any $S \subseteq N$ we let V_S denote the subspace spanned by $\{a^j : j \in S\}$, and we let $P_S(b)$ denote the projection of b in V_S. That is, $P_S(b) = \sum_{j \in S} a^j x_j$ where x solves the least squares problem $\min_{x \in \Re^S} \left\| \sum_{j \in S} a^j x_j - b \right\|^2$. Throughout the paper, we let $\|\cdot\|$ and $\langle \cdot, \cdot \rangle$ denote the Euclidean norm and inner product, respectively, and we assume that the target and the atoms are normalized, i.e. $\|a^1\| = \cdots = \|a^n\| = \|b\| = 1$. We let I denote the identity matrix and e^i the ith unit vector which dimensions are consistent with the context. If v is a vector, we let v_i denote the ith component of v.

The goal is to find an $S \subseteq N$, where $|S| \leq t$, such that the distance between b and $P_S(b)$ is minimized, and hence the (Euclidean) *Subspace Selection Problem* can be defined as

$$\min_{S \subseteq N} \{\|P_S(b) - b\| : |S| \leq t\} \quad [\textbf{SSP}] \ .$$

The problem is frequently referred to as the *subset selection problem*, but is rephrased here for better distinction from other combinatorial optimization problems consisting in subset selection. By a reduction from *Exact Cover by 3-sets* (Garey and Johnson [8], pp. 221), Natarajan [6] proved that [**SSP**] is NP-hard. This has drawn the attention to various heuristic approaches, to be discussed briefly as follows.

The Matching Pursuit algorithm [3], relies on estimations $\tilde{b}$ of the projection $P_S(b)$. In each iteration, the algorithm picks the $j \in N$ maximizing $\left| \left\langle b - \tilde{b}, a^j \right\rangle \right|$, and updates its weight. As it may happen that the maximum is attained for some $j \in S$, the method does not necessarily extend S in each iteration, and examples can be found (see Chen et al. [9]) where the method does not converge.

The shortcoming of Matching Pursuit is to some extent accounted for in the Orthogonal Matching Pursuit (OMP) algorithm [4]. In each iteration, the residual $b - P_S(b)$ is computed by solving the associated least-squares problem. Then $|\langle a^j, b - P_S(b) \rangle|$ is maximized with respect to $j \in N \setminus S$ in order to find the entering atom. Hence finite convergence is guaranteed.

The selection criterion in OMP does not necessarily result in the best extension by one. The forward greedy heuristic [6], or Fully Orthogonal Matching Pursuit [5], is identical to OMP, except that the selection criterion is modified such that it corresponds to an optimal extension by one. Details on this are given in the next section.

The backward greedy algorithm of Couvreur and Bresler [2] contrasts the forward methods by initially accepting all atoms, and next eliminate one in each of $n-t$ iterations. The elimination criterion corresponds to an optimal reduction by one. An important prerequisite for this approach is that $n \leq m$, i.e. only non-overcomplete dictionaries can be handled. The reason for this is that in the case where A is overcomplete, any column set the deletion of which leaves A with full rank, can be eliminated. Hence the first $n - m$ iterations give no increase in the objective function, and the elimination criterion is ambiguous as it gives no indication of what atoms to leave out.

Under the assumption that $n \leq m$, it is proved in [2] that the backward greedy algorithm is exact in cases where the optimal objective function value is sufficiently small. More precisely, if b is within an ε-neighborhood of a linear combination of no more than t atoms, the algorithm returns exactly this combination. Computing the bound ε is however conjectured to be no easier than solving the problem.

2.1 Basic Concepts from Numerical Linear Algebra

For readers unfamiliar with the matrix computation operations used in this work, we summarize them briefly in this section. Details can be found in e.g. the book of Golub and Van Loan [10].

Let $(j_1, \ldots, j_n)$ be a permutation of N, and let Π be the corresponding column permutation matrix. If $Q \in \Re^{m \times m}$ and $R = (r_{ij})$ such that $QR = A\Pi$, $Q^T Q = I$ and for some non-negative integer $k \leq m$ we have $r_{ij} = 0 \; \forall j = 1, \ldots, k; i = j + 1, \ldots, m$, then we refer to QR as a *partial QR-decomposition* of $A\Pi$. The objective function value corresponding to the selection of atoms $S = \{j_1, \ldots, j_k\}$ is found as the sum of the squares of the $m - k$ last components of $Q^T b$. The first k columns of Q are an orthonormal basis for the selected subspace V_S.

If $v \in \Re^m$ and $H_k(v) = I - \frac{2}{u^T u} u u^T$, where

$$u = \left(0, \ldots, 0, v_k \pm \sqrt{\textstyle\sum_{i=k}^m v_i^2}, v_{k+1}, \ldots, v_m\right)^T \in \Re^m,$$ then $H_k(v)$ is said to be a *Householder reflection*. It is readily seen that $H_k(v) = H_k^T(v)$ and $H_k^T(v) H_k(v) = I$, and that $H_k(v)$ maps v to a vector with zeros in positions $k + 1, \ldots, m$. More precisely, $H_k(v)v = w$, where $w_1 = v_1, \ldots, w_{k-1} = v_{k-1}, w_k = \mp \sqrt{\sum_{i=k}^m v_i^2}$, and $w_{k+1} = \cdots = w_m = 0$.

Hence partial QR-decompositions of $A\Pi$ are obtained by applications of Householder reflections. By careful selections of vectors $v^1, \ldots, v^k$, we get zeros below the diagonal in the k first columns of $H_k(v^k) \cdots H_1(v^1) A\Pi = R$, and thereby $Q^T = H_k(v^k) \cdots H_1(v^1)$, which means that $Q = H_1(v^1) \cdots H_k(v^k)$. Actually, v^1 is set equal to a^{j_1}, v^2 is set to $H_1(v^1) a^{j_2}$, etc.

A matrix $G_{ik}(v)$, which is identical to the identity matrix except that it has the value $\frac{v_i}{\sqrt{v_i^2 + v_k^2}}$ in positions (i, i) and (k, k), and the values $\frac{v_k}{\sqrt{v_i^2 + v_k^2}}$ and $\frac{-v_k}{\sqrt{v_i^2 + v_k^2}}$ in positions (i, k) and (k, i), respectively, is referred to as a *Givens rotation*. Clearly, $G_{ik}(v) = -G_{ik}^T(v)$ and $G_{ik}^T(v) G_{ik}(v) = I$. If $w = G_{ik}(v)v$, then $w_j = v_j$ $\forall j \in \{1, \ldots, m\} \setminus \{i, k\}$, $w_i = \sqrt{v_i^2 + v_k^2}$, and $w_k = 0$. Hence $G_{ik}(v)$ transforms v to a vector with a zero in position k.

A matrix $R \in \Re^{m \times n}$ is an *upper Hessenberg* matrix if $r_{ij} = 0 \; \forall i = j + 2, \ldots, m; j = 1, \ldots, \min\{n, m - 2\}$, which means that R has zeros below its lower subdiagonal. A set of Givens rotations are useful for transforming an upper Hessenberg matrix to an upper triangular matrix.

3   Extension and Reduction Neighborhoods

Any feasible solution to [**SSP**] is identified by a set of atoms $S = \{j_1, \ldots, j_k\} \subseteq N$. Associated with such solutions, our heuristic considers the neighborhoods $\mathcal{N}^+(S) = \{S^+ = S \cup \{j\}, j \in N \setminus S\}$ and $\mathcal{N}^-(S) = \{S^- = S \setminus \{j\}, j \in S\}$ and the corresponding subproblems defining best moves:

$$\min_{j \in N \setminus S} \left\| b - P_{S \cup \{j\}}(b) \right\| \qquad [SSP^+(\mathbf{S})]$$

$$\min_{j \in S} \left\| b - P_{S \setminus \{j\}}(b) \right\| \qquad [SSP^-(\mathbf{S})]$$

Note that the union of the neighborhoods actually corresponds to the single-bit-flip neighborhood (the Hamming distance 1 neighborhood). Conditions useful for identifying the best moves are given next.

3.1   Extending the Solution

Assume the partial QR-decomposition $QR = A\Pi$, where $Q = (q^1, \ldots, q^m)$, $R = (r^{j_1}, \ldots, r^{j_n})$, and the associated residual $\rho_S = b - P_S(b)$.

Proposition 1. *If $k < m$, then $j^+ \in N \setminus S$ is an optimal solution to $[SSP^+(\mathbf{S})]$ if and only if*

$$j^+ \in \arg \max_{j \in N \setminus S} \frac{\langle \rho_S, a^j \rangle^2}{1 - \sum_{\ell=1}^{k} \langle q^\ell, a^j \rangle^2} \, . \qquad (1)$$

Proof. Since $\{q^1, \ldots, q^k\}$ is an orthonormal basis for V_S, and $k < m$, there exists for all $j \in N \setminus S$ some $q(j) \in \Re^m$ such that $\{q^1, \ldots, q^k, q(j)\}$ is an orthonormal basis for $V_{S \cup \{j\}}$. We have that $\left\| \rho_{S \cup \{j\}} \right\| = \left\| b - P_{S \cup \{j\}}(b) \right\|^2 = \|b\|^2 - \sum_{\ell=1}^{k} \langle b, q^\ell \rangle^2 - \langle b, q(j) \rangle^2$, which is minimized if and only if j is chosen such that $\langle b, q(j) \rangle^2$ is maximized. But

$$a^j = \langle a^j, q(j) \rangle q(j) + \sum_{\ell=1}^{k} \langle a^j, q^\ell \rangle q^\ell,$$

and since $\left\| a^j \right\| = 1$, we get by taking the inner product of b and the above equation

$$\langle b, q(j) \rangle^2 = \frac{\left(\langle b, a^j \rangle - \sum_{\ell=1}^{k} \langle a^j, q^\ell \rangle \langle b, q^\ell \rangle \right)^2}{\langle a^j, q(j) \rangle^2} = \frac{\left\langle a^j, b - \sum_{\ell=1}^{k} \langle b, q^\ell \rangle q^\ell \right\rangle^2}{\langle a^j, q(j) \rangle^2}$$

$$= \frac{\langle a^j, \rho_S \rangle^2}{1 - \sum_{\ell=1}^{k} \langle a^j, q^\ell \rangle^2} \, .$$

□

In the case of a normalized dictionary, this is precisely the expansion criterion applied in the greedy algorithm in [6], although no proof of its optimality was given. In [4], however, the enumerator of (1) is maximized.

When adding j^+ to S, we have to update the QR-decomposition. This is done by applying a Householder reflection, exactly as in a single iteration of the QR-algorithm for the linear least-squares problem (see e.g. [10]).

Since $\langle q^\ell, a^j \rangle = r_\ell^j$, the expansion criterion (1) suggests that the algorithm stores ρ_S and R. The Householder reflection in question zeros out all the $m-k-1$ last entries of $Q^T a^{j^+}$, while the first k entries remain unchanged. Once the entering j^+ is found, we compute the corresponding reflection, H, and overwrite R by applying H to all columns r^j where $j \in N \setminus S$. The reflection does not alter $r^{j_1}, \ldots, r^{j_k}$.

Let $\alpha = Q^T b$, and note that

$$P_S(b) = \sum_{\ell=1}^{k} \langle b, q^\ell \rangle q^\ell = \sum_{\ell=1}^{k} \alpha_\ell q^\ell \tag{2}$$

and

$$\rho_S = \sum_{\ell=k+1}^{m} \langle b, q^\ell \rangle q^\ell = \sum_{\ell=k+1}^{m} \alpha_\ell q^\ell \, . \tag{3}$$

By the initial assignments $Q = I$ and $\alpha = b$, the representations (2)-(3) can be maintained by applying H also to α. Because $\langle q^\ell, a^j \rangle = r_\ell^j$ and $\langle \rho_S, a^j \rangle = \sum_{\ell=k+1}^{m} r_\ell^j \alpha_\ell$, the best word to enter is computed through

$$j^+ \in \arg \max_{j \in N \setminus S} \frac{\left(\sum_{\ell=k+1}^{m} r_\ell^j \alpha_\ell \right)^2}{1 - \sum_{\ell=1}^{k} \left(r_\ell^j \right)^2} \, . \tag{4}$$

When moving forward, we shall also maintain some variables that turn out to facilitate the backward move. Reeves [11] has shown (see the next section) that the matrix $Y = (y^1, \ldots, y^k) = \begin{pmatrix} \tilde{R}^{-T} \\ 0 \end{pmatrix} \in \Re^{m \times k}$, where $\tilde{R}$ denotes the upper triangular submatrix consisting of the elements in the first k rows and columns of R, is useful for this purpose. The superscript $-T$ means taking the transpose of the inverse. Also the weights $x \in \Re^k$ satisfying $P_S(b) = \sum_{\ell=1}^{k} a^{j_\ell} x_\ell$ will be computed for the purpose of the backward move.

Since the first k columns of $\tilde{R}$ are not affected by inclusion of the $(k+1)$st atom, we do not have to recompute the k first rows of Y either. Row $k+1$ is found by simply solving a triangular linear system of equations. Finally, it is easy to show [10] that $x = Y^T \alpha$.

Algorithm 1. forward($\underline{\delta}$, S, z)

$k \leftarrow |S|$

find a j^+ satisfying (4)

if $\left(\sum_{\ell=k+1}^{m} r_\ell^{j^+} \alpha_\ell\right)^2 \leq \underline{\delta} \left(1 - \sum_{\ell=1}^{k} \left(r_l^{j^+}\right)^2\right)$ **then**

    return **false**

$k \leftarrow k+1, j_k \leftarrow j^+, S \leftarrow S \cup \{j_k\}$

$H \leftarrow H_k\left(r^{j_k}\right)$

$R \leftarrow HR$, $\alpha \leftarrow H\alpha$

$\tilde{R} \leftarrow$ triangular submatrix of R

$Y \leftarrow (Y, 0)$, solve $\tilde{R}u = e^k$ for u, and let the kth row of Y be u^T

$x \leftarrow Y^T \alpha$

$z \leftarrow \sum_{\ell=k+1}^{m} \alpha_\ell^2$

return **true**

Algorithm 1. performs a forward greedy move, along with all necessary updates. For reasons to be made clear in Sect. 4, we equip the algorithm with the ability to reject a forward move if the objective function reduction is below a given threshold $\underline{\delta}$. Input parameters are $\underline{\delta}$ and the current solution S, and the updated solution S and its corresponding objective function value z are output via the parameter list. A Boolean value telling whether or not a move was executed is returned. We assume that all other algorithms are notified about assignments to variables α, x, Y, and R.

Construction of H and the application of it to a single column takes $\mathcal{O}(m)$ time, and the reflection is applied to $\mathcal{O}(n)$ columns. Solving the triangular equation system takes $\mathcal{O}\left(k^2\right)$ time. Since $k \leq \min\{m, n\}$, the time complexity of Algorithm 1. is thus $\mathcal{O}(mn)$. Some computations may be made faster than shown in Algorithm 1., but as long as the running time remains quadratic, we have chosen not to present details on this.

3.2   Reducing the Solution

Note that the entering j^+ is found without explicitly computing the new QR-decomposition. Due to Reeves [11], this is also the case when the best reduction of S is to be found.

Proposition 2. *If $k < m$, then $j^- \in N \setminus S$ is an optimal solution to $[SSP^-(\mathbf{S})]$ if and only if*

$$j^- \in \arg\min_{j_\ell \in S} \frac{x_\ell^2}{\|y^\ell\|^2} \ . \tag{5}$$

Proof. See [11].   □

Actually, the expression to be minimized in (5) corresponds to the increase in the objective function value implied by the elimination of j_ℓ.

The matrix $\left(r^{j_1}, \ldots, r^{j_{\ell-1}}, r^{j_{\ell+1}}, \ldots, r^{j_k}\right)$ is upper Hessenberg with non-zero subdiagonal elements in columns $\ell+1, \ldots, k$. By applications of $k - \ell$ Givens

---

Algorithm 2. backward($\bar{\delta}$, S, z)

$k \leftarrow |S|$

find $\ell^- \in \arg\min_{\ell \in \{1,\ldots,k\}} \left\{ \frac{x_\ell^2}{\|y^\ell\|} \right\}$

if $x_{\ell^-}^2 \geq \bar{\delta} \left\| y^{\ell^-} \right\|^2$ **then**
    return **false**
$k \leftarrow k - 1, S \leftarrow S \setminus \{j_{\ell^-}\}$
for $\ell \leftarrow \ell^-, \ell^- + 1, \ldots, k$ **do**
    $j_\ell \leftarrow j_{\ell+1}, y^\ell \leftarrow y^{\ell+1}$
$Y \leftarrow (y^1, \ldots, y^k)$
for $\ell \leftarrow \ell^-, \ell^- + 1, \ldots, k$ **do**
    $G \leftarrow$ the Givens rotation $G_{\ell,\ell+1}\left(r^{j_\ell}\right)$
    $R \leftarrow GR, \alpha \leftarrow G\alpha, Y \leftarrow GY$
$x \leftarrow Y^T \alpha$
$z \leftarrow \sum_{\ell=k+1}^{m} \alpha_\ell^2$
return **true**

---

rotations, the first of which is $G_{\ell,\ell+1}\left(r^{j_{\ell+1}}\right)$, a partial QR-decomposition of $A\Pi'$, where Π' is a permutation matrix putting $a^{j_1}, \ldots, a^{j_{\ell-1}}, a^{j_{\ell+1}}, \ldots, a^{j_k}$ in the first $k - 1$ positions, is established. We also apply the rotations to the columns of Y, which ensures that the upper left $(k - 1) \times (k - 1)$ submatrix of Y remains an inverse of $\tilde{R}^T$. Hence the backward greedy move is as given in Algorithm 2. If the increase in the objective function value is as large as the input argument $\bar{\delta}$, we reject the move.

In the worst case, the atom to leave S is j_1, and we have to apply $k - 1$ rotations. Since the construction and the application of a Givens rotation matrix takes constant time, it follows from $k \leq \min\{m, n\}$ that the running time of Algorithm 2. is $\mathcal{O}(mn)$.

4   Bidirectional Greedy Search

All forward and backward greedy heuristics suffer from lack of ability to back up if unfortunate moves are made early in the process. Bad effects of the greedy methods could be reduced by combining them with search in other neighborhoods than $\mathcal{N}^+(S)$ and $\mathcal{N}^-(S)$. For instance, if a local optimal solution with respect to the exchange by one neighborhood is computed immediately after the addition/elimination of an atom, it is likely that significantly better solutions could be located.

In the case of [**SSP**], the size of the exchange by one neighborhood is $|S|(n - |S|)$, and we are aware of no method for evaluating a single neighbor without virtually performing both of Algorithms 1. and 2.. Since a neighborhood evaluation procedure with time complexity of the fourth order is undesirable, we rather suggest a search that entirely is based on $\mathcal{N}^+(S)$ and $\mathcal{N}^-(S)$. This results in a *bidirectional* greedy heuristic of which we study two versions.

The first version can be seen as a modification of the forward greedy heuristic in that it starts with $S = \emptyset$, and has forward moves as its primary operation. Throughout the search, it keeps track of $S_1, \ldots, S_t$, where S_k is the best known solution with cardinality k. When $|S| = k$, and a backward greedy move can result in a solution better than S_{k-1}, this move is preferred. Otherwise, a forward move is tried unless $|S| = t$, in which case the algorithm stops. Also the forward move is rejected if there is no solution in $\mathcal{N}^+(S)$ better than S_{k+1}. In that case, a deterministic move, referred to as a jump, to the solution S_{k+1} is made. Thus finite convergence, to a possibly suboptimal solution, is ensured. The heuristic is given in Algorithm 3.

A backward version of this heuristic is valid in the case $n \leq m$. Then we start with $S = N$, and in each iteration we move to the best solution in $\mathcal{N}^+(S)$ if it is better than S_{k+1}. Otherwise we move to the best of the best solution in $\mathcal{N}^-(S)$ and S_{k-1}, unless the stopping criterion $|S| = t$ is satisfied. Hence, with only minor changes, the algorithmic steps are the same as those of Algorithm 3.

For both heuristics, the jumps are accomplished by a sequence of moves in the two neighborhoods. Each jump implies $|S_{k\pm1} \setminus S|$ and $|S \setminus S_{k\pm1}|$ forward and backward moves, respectively.

The bidirectional greedy algorithm resembles *Variable Neighborhood Search* (VNS) introduced by Mladenović and Hansen [12] in the sense that more than one neighborhood is considered. Another point of resemblance is that if the search in one neighborhood is unsuccessful, search is directed to the next. However, our approach is distinguished from VNS in that the solution found in $\mathcal{N}^+(S)$ or $\mathcal{N}^-(S)$ may be accepted even if it is no better than the currently best. In fact, there is no solution in $\mathcal{N}^-(S)$ with smaller objective function value than S, and consequently the definition of success of the search in this neighborhood must be more liberal.

4.1   Randomized Search

Both versions of the bidirectional greedy algorithm can be modified in order to limit the number of moves opposite to the primary search direction. This will lead to faster convergence, but at the expected expense of weaker solution quality. One way of controlling the extent of reverse moves, is to draw randomly if a beneficial secondary move should be approved. By introducing a fixed approval probability, p, we have the deterministic variants of the bidirectional greedy methods if $p = 1$, and the standard one-directional greedy methods if $p = 0$. Any value in the range $(0, 1)$ reflects the trade-off between computational speed (small p) and solution quality (large p).

5   Numerical Experiments

We have implemented both versions of our bidirectional greedy heuristic, with both deterministic and randomized search. All code is written in C++ and compiled by the GCC 4.1.1 compiler, and all experiments reported are carried out on a Pentium M 1.73 GHz computer with a Fedora Linux operating system.

Algorithm 3. `bigreedyforward`(S, z^*)

$S \leftarrow \emptyset, z^* \leftarrow \infty$

$z_1 \leftarrow z_2 \leftarrow \cdots \leftarrow z_t \leftarrow \infty$

repeat

    $k \leftarrow |S|$

    `move` $\leftarrow$ `false`

    if $k > 2$ **then**

        `move` $\leftarrow$ `backward`$(z_{k-1} - z^*, S, z^*)$

    if `move` **then**

        $S_{k-1} \leftarrow S, z_{k-1} \leftarrow z^*$

    else if $k < t$ **then**

        `move` $\leftarrow$ `forward`$(z^* - z_{k+1}, S, z^*)$

        if `move` **then**

            $S_{k+1} \leftarrow S, z_{k+1} \leftarrow z^*$

        else

            $S \leftarrow S_{k+1}, z^* \leftarrow z_{k+1}$

    else

        `done` $\leftarrow$ `true`

until `done`

5.1 Forward Search

In order to study the performance of our methods, we first apply them to a set of randomly generated instances of size $m = 20$, $n = 100$, and where the subspace dimension is set to $t = 2, 4, \ldots, 10$. For each such triple (m, n, t) we generate 100 instances by first drawing all dictionary and target elements from a uniform distribution on $[-1, 1]$, and then normalize such that $\|b\| = 1$ and $\|a^j\| = 1$ for all $j \in N$. Each instance is solved by the forward version of the bidirectional greedy algorithm, for each of the approval probabilities 0.00, 0.25, 0.50, 0.75 and 1.00.

In Table 1, we give the objective function values [`solution`], that is the squared residual norm, averaged over all 100 instances within a set. A column in the table corresponds to an approval probability. In the same table, we also give the number of secondary (backward) moves performed [`moves`]. This includes the $|S \setminus S_{k+1}|$ backward moves required when a jump is made. Note that for the forward version of our algorithm, the total number of moves is given by t plus twice the number of reverse moves.

Table 1 shows as expected that the bidirectional approach gives better results than the pure forward greedy method. The largest improvement is for the largest value of t, in which case the reduction in objective function value is 19%. This comes at the computational price of 11.82 moves in average, 0.91 of which are backward, or an increase of 18.2% from the 10 forward moves of the pure greedy method. For smaller values of t, the effect is smaller. Throughout we see that putting the approval probabilities between 0 and 1 for most instance sets gives results between those obtained for the extreme values of this parameter. The

Table 1. Average results for 100 instances where $m = 20$ and $n = 100$

	$p = 0.00$	$p = 0.25$		$p = 0.50$		$p = 0.75$		$p = 1.00$	
t	solution	solution	moves	solution	moves	solution	moves	solution	moves
2	0.43905	0.43905	0.00	0.43905	0.00	0.43905	0.00	0.43905	0.00
4	0.19534	0.19315	0.07	0.19388	0.04	0.19282	0.11	0.19275	0.12
6	0.08194	0.08102	0.06	0.07916	0.21	0.07911	0.24	0.07862	0.32
8	0.02975	0.02809	0.19	0.02793	0.31	0.02723	0.49	0.02643	0.63
10	0.00854	0.00753	0.36	0.00756	0.46	0.00722	0.82	0.00692	0.91

solution quality tends to improve with increasing value of p, but for $p = 0.5$, there are two exceptions to this rule ($t = 4$ and $t = 10$).

A similar experiment where $m = 100$ and $n = 1000$ is reported in Table 2. Here we let t grow beyond $m/2$, although this is not likely to be the case for the application described in the introduction. For other applications [1], we may however have $t \geq m/2$. We observe the same effects, but when t is sufficiently large, any value of p seems to give a subspace where the target can be approximated with very small error. The best relative improvement over the one-directional greedy method is obtained for $t = 50$ (28% reduction, 63% more moves).

Table 2. Average results for 100 instances where $m = 100$ and $n = 1000$

	$p = 0.00$	$p = 0.25$		$p = 0.50$		$p = 0.75$		$p = 1.00$	
t	solution	solution	moves	solution	moves	solution	moves	solution	moves
10	0.34120	0.33892	0.36	0.33748	0.63	0.33459	1.21	0.33396	1.69
20	0.11353	0.10988	1.49	0.10743	2.99	0.10625	4.33	0.10460	5.57
30	0.03249	0.03037	2.75	0.02901	5.00	0.02811	7.25	0.02759	9.62
40	0.00752	0.00661	3.34	0.00621	7.23	0.00594	10.31	0.00587	13.10
50	0.00123	0.00098	4.29	0.00094	8.27	0.00091	11.73	0.00088	15.82
60	0.00011	0.00010	4.70	0.00009	10.18	0.00008	13.63	0.00008	17.73

5.2 Comparing Forward and Backward Search

In order to compare the forward bidirectional greedy approach with the backward version, we also study two instance sets where $m = n$. As above, we let m attain the values 20 and 100, and we let t vary from 4 to 16 and from 10 to 90, respectively. For $p \in \{0.00, 0.25, 0.50, 0.75, 1.00\}$, we apply both bidirectional heuristics to 100 instances in each of the two sets. Results are given in Tables 3-4.

Comparing the one-directional methods first, we see that the forward version gives better results for values of t as large as 12 and 80 in the two instance sets. It is not surprising that the competitiveness of backward greedy grows with subset size, but it is noteworthy that the forward version gives superior results even for $t > m/2$. This observation is also made for the bidirectional heuristic, although putting $p = 1$ helps the backward version to get closer to the forward one (for

Table 3. Average results for 100 instances where $m = n = 20$

t	$p = 0.00$ solution	$p = 0.25$ solution	moves	$p = 0.50$ solution	moves	$p = 0.75$ solution	moves	$p = 1.00$ solution	moves
				primary direction forward					
4	0.45215	0.44862	0.10	0.44857	0.15	0.44700	0.22	0.44597	0.25
8	0.24565	0.23658	0.51	0.22823	1.22	0.22595	1.46	0.22447	1.48
12	0.12903	0.12136	1.09	0.11213	2.53	0.10715	3.62	0.10498	4.10
16	0.05980	0.05475	1.99	0.05023	4.42	0.04725	6.82	0.04404	8.19
				primary direction backward					
4	0.55158	0.49251	2.16	0.46355	5.64	0.46577	8.18	0.45447	10.15
8	0.30559	0.28097	1.45	0.25125	3.20	0.24270	5.17	0.23056	6.58
12	0.14160	0.13291	0.64	0.12541	1.24	0.12202	1.86	0.11303	2.52
16	0.04046	0.03911	0.09	0.03684	0.24	0.03360	0.46	0.03356	0.55

Table 4. Average results for 100 instances where $m = n = 100$

t	$p = 0.00$ solution	$p = 0.25$ solution	moves	$p = 0.50$ solution	moves	$p = 0.75$ solution	moves	$p = 1.00$ solution	moves
				primary direction forward					
10	0.59259	0.59026	0.71	0.58905	1.17	0.58837	1.68	0.58754	2.02
20	0.40824	0.40054	2.22	0.39886	4.53	0.39436	8.23	0.39285	10.11
30	0.29254	0.27973	4.86	0.27337	10.74	0.26988	18.06	0.27003	23.05
40	0.21154	0.19714	7.61	0.18889	17.88	0.18514	30.39	0.18314	39.86
50	0.15202	0.13553	11.53	0.12896	27.53	0.12408	45.82	0.12305	62.52
60	0.10830	0.09361	15.19	0.08223	37.41	0.07959	67.51	0.07805	90.85
70	0.07346	0.06072	19.35	0.05179	48.87	0.04759	90.76	0.04532	125.63
80	0.04581	0.03573	24.96	0.02833	62.70	0.02590	114.81	0.02473	162.05
90	0.02300	0.01712	28.49	0.01242	83.87	0.01101	154.18	0.00977	214.27
				primary direction backward					
10	0.67832	0.60670	25.54	0.59562	69.35	0.59290	160.61	0.59119	231.54
20	0.52567	0.42805	23.84	0.40912	66.07	0.39911	140.76	0.39863	210.57
30	0.41227	0.30971	20.17	0.28667	57.61	0.27784	130.56	0.27605	182.22
40	0.32013	0.23273	16.74	0.20066	48.65	0.19396	105.47	0.18852	149.91
50	0.23876	0.16640	13.49	0.14455	37.16	0.13405	79.14	0.13014	112.86
60	0.16654	0.12318	8.86	0.09887	27.99	0.09207	51.30	0.08708	78.19
70	0.10153	0.07433	5.99	0.06209	17.56	0.05639	31.36	0.05412	42.78
80	0.04899	0.04092	2.90	0.03491	7.23	0.02972	14.98	0.02629	20.91
90	0.01289	0.01193	0.76	0.01088	1.80	0.01026	2.74	0.00903	3.99

$t = 12$ and $t = 80$, the relative superiority of the forward version is reduced from 8.9% to 7.1% and from 6.5% to 5.9%, respectively).

Comparing the work load of the two versions of bidirectional search with $p = 1$, we see that for $t = 12$ in the first instance set the forward version requires 20.20 moves in average, whereas the backward version takes 20 forward moves for initiating S to N (compute the QR-factorization of A), and then in average 10.52 backward and 2.52 forward moves. Hence the better results obtained by the forward version come at no higher computational cost. In the second instance

Table 5. Average performance for 100 instances where $m = n = 100$

t	$p = 0.00$ tmoves	cpu	$p = 0.25$ tmoves	cpu	$p = 0.50$ tmoves	cpu	$p = 0.75$ tmoves	cpu	$p = 1.00$ tmoves	cpu
				primary direction forward						
10	10.00	0.00	11.42	0.00	12.34	0.00	13.36	0.00	14.04	0.00
20	20.00	0.01	24.44	0.01	29.06	0.01	36.46	0.01	40.22	0.02
30	30.00	0.01	39.72	0.01	51.48	0.02	66.12	0.03	76.10	0.04
40	40.00	0.01	55.22	0.03	75.76	0.05	100.78	0.08	119.72	0.11
50	50.00	0.01	73.06	0.05	105.06	0.10	141.64	0.18	175.04	0.26
60	60.00	0.02	90.38	0.08	134.82	0.19	195.02	0.37	241.70	0.56
70	70.00	0.02	108.70	0.12	167.74	0.34	251.52	0.69	321.26	1.11
80	80.00	0.02	129.92	0.20	205.40	0.58	309.62	1.23	404.10	1.97
90	90.00	0.03	146.98	0.29	257.74	1.00	398.36	2.23	518.54	3.63
				primary direction backward						
10	190.00	0.83	241.08	1.00	328.70	1.33	511.22	2.04	653.08	2.59
20	180.00	0.82	227.68	1.01	312.14	1.33	461.52	1.99	601.14	2.55
30	170.00	0.81	210.34	1.00	285.22	1.32	431.12	2.00	534.44	2.49
40	160.00	0.79	193.48	0.97	257.30	1.29	370.94	1.90	459.82	2.36
50	150.00	0.76	176.98	0.92	224.32	1.20	308.28	1.72	375.72	2.13
60	140.00	0.70	157.72	0.82	195.98	1.09	242.60	1.44	296.38	1.82
70	130.00	0.60	141.98	0.71	165.12	0.90	192.72	1.14	215.56	1.35
80	120.00	0.47	125.80	0.53	134.46	0.62	149.96	0.77	161.82	0.90
90	110.00	0.28	111.52	0.30	113.60	0.33	115.48	0.35	117.98	0.38

set, this is no longer true, as the average number of moves for $t = 80$ are 404.10 and 161.82, respectively. Only for $t \leq 60$ is the number of moves smaller in the forward version.

To compare the work load further, we have also given the running time in CPU seconds [cpu] for the last instance set in Table 5. In comparison, we give the total number of moves [tmoves], including moves in both the primary and secondary direction. Hence the move counts reported in Table 5 equal twice those of Table 4, plus t when the primary direction is forward, and plus $2n - t$ otherwise.

When the primary direction is forward (backward), we see, by reading Table 5 vertically, that the running time increases slightly faster than the number of moves with increasing (decreasing) t. A similar observation is made when reading horizontally, i.e. when t is constant and p increases. The running times confirm that the forward version is the more efficient for $t \leq 60$. When the number of moves are approximately equal, e.g. when $t = 70$ and $p = 0.50$, the forward version is faster. This is explained by the fact that in all but the first t moves of the backward version, we have $k > t$ in Algorithms 1.-2., whereas $k \leq t$ in all moves of the forward version. In both algorithms, a triangular system of k linear equations is solved, requiring $\mathcal{O}\left(k^2\right)$ time.

Finally, we note that the backward version has the best worst-case running time for $m = n = 100$. The instances in its hardest subset ($t = 10$, $p = 1.00$) could be solved in averagely 2.59 seconds, whereas the forward version needed 3.63 seconds in average for its hardest instances ($t = 90$, $p = 1.00$).

6   Conclusion

We have studied the problem of finding the best subspace of low dimension for projecting a given target vector. The selection criterion is that the projection, in a least squares sense, is to be as close as possible to the target. A set of possible basis vectors for the subspace are given, and the problem is to choose a given number of these.

We demonstrate that the moves in the forward and greedy heuristics of this NP-hard problem can be carried out in quadratic time. Conversely, the exchange by one move seems to require two orders of magnitude more computations. Motivated by these observations, we suggest a generalized greedy approach where the search alternates between forward and backward greedy moves (extensions and reductions by one). This results in a bidirectional greedy algorithm with either a forward or a backward primary direction. The algorithm prefers a move opposite to its primary direction whenever this produces a solution better than the best known solution of the resulting cardinality. We have also given a randomized extension of the heuristic.

Computational tests show that the suggested bidirectional greedy methods are superior to their one-directional counterparts. Furthermore, the forward version seems to give the best results unless the subspace dimension is significantly higher than half the dimension of the target.

The validity of the approach taken in this work is not restricted to the particular problem under study. An interesting topic for future research is to identify other combinatorial optimization problems where this approach can give significant improvements over greedy search in one direction.

References

1. Miller, A.J.: Subset Selection in Regression, 2nd edn. Chapman and Hall, London, U.K (2002)
2. Couvreur, C., Bresler, Y.: On the optimality of the backward greedy algorithm for the subset selection problem. SIAM Journal on Matrix Analysis and Applications 21, 797–808 (2000)
3. Mallat, S., Zhang, Z.: Matching Pursuit in a Time-frequency Dictionary. IEEE Transactions on Signal Processing 41, 3397–3415 (1993)
4. Pati, Y.C., Rezaiifar, R., Krishnaprasad, P.S.: Orthogonal Matching Pursuit: Recursive function approximation with applications to wavelet decomposition. In: Proc. 27th Annu Asilomar Conf. Signals, Systems and Computers, Pacific Grove, CA, pp. 40–44 (1993)
5. Gharavi-Alkhansari, M., Huang, T.S.: A Fast Orthogonal Matching Pursuit Algorithm. In: Proc. ICASSP '98, Seattle, Washington, USA, pp. 1389–1392 (1998)
6. Natarajan, B.K: Sparse approximate solutions to linear systems. SIAM J Comput. 24, 227–234 (1995)
7. Haugland, D., Storøy, S.: Local search methods for ℓ_1-minimization in frame based signal compression. Optimization and Engineering 7, 81–96 (2006)
8. Garey, M.R., Johnson, D.S.: Computers and Intractability: A Guide to the Theory of NP-completeness. W.H. Freeman, New York (1979)

9. Chen, S.S., Donoho, D.L., Saunders, M.A.: Atomic decomposition by Basis Pursuit. SIAM Journal on Scientific Computing 20, 33–61 (1998)
10. Golub, G.H., Van Loan, C.F.: Matrix Computations, 3rd edn. Johns Hopkins Univ. Press, Baltimore, MD (1996)
11. Reeves, S.J.: An Efficient Implementation of the Backward Greedy Algorithm for Sparse Signal Reconstruction. IEEE Signal Processing Letters 6, 266–268 (1999)
12. Mladenović, N., Hansen, P.: Variable Neighborhood Search. Computers and Operations Research 24, 1097–1100 (1997)

EasySyn++: A Tool for Automatic Synthesis of Stochastic Local Search Algorithms

Luca Di Gaspero and Andrea Schaerf

DIEGM, University of Udine, Udine, Italy
{l.digaspero,schaerf}@uniud.it

Abstract. We present a software tool, called EASYSYN++, for the automatic synthesis of the source code for a set of stochastic local search (SLS) algorithms. EASYSYN++ uses C++ as object language and relies on EASYLOCAL++, a C++ framework for the development of SLS algorithms. EASYSYN++ is particularly suitable for the frequent case of having many neighborhood relations that are potentially useful.

1 Introduction

In this work we present a software tool, called EASYSYN++, for the automatic synthesis of the source code for a set of local search algorithms. The synthesized algorithms range from basic local search methods, like hill climbing, simulated annealing, and tabu search, to complex strategies, such as variable neighborhood search and iterated local search.

The tool is implemented in PHP and it uses C++ as object language. EASY-SYN++ relies on the C++ local search framework, EASYLOCAL++ [1,2], which has recently been entirely redesigned, and on EASYANALYZER [3] for the experimental analysis of SLS algorithms.

EASYSYN++ input is an XML description of the basic building blocks of local search, namely the search space, the neighborhood relation(s), and the cost function components (hard and soft) for the specific problem. Its output is a set of C++ classes that have to be completed with some problem-specific code developed by the user, and have to be compiled against the problem-independent abstract classes of EASYLOCAL++.

EASYSYN++ is particularly useful for the frequent case of many neighborhood relations (see, e.g., [4]). For these cases, EASYSYN++ manages automatically the composition of basic neighborhoods defining more complex ones. As a consequence, the set of possible combinations of algorithms is relatively large, and the human coding of the whole set of potential algorithms would be time consuming and prone to errors and inconsistencies.

Needless to say, EASYSYN++ cannot replace the human experience in designing the full-fledged algorithm with all its peculiarities. Nevertheless, we believe that this tool can help in a preliminary exploratory phase in which many alternatives are evaluated before focusing on the most promising ones. Thus, EASY-LOCAL++, EASYANALYZER, and EASYSYN++ allow the user to obtain with

T. Stützle, M. Birattari, and H.H. Hoos (Eds.): SLS 2007, LNCS 4638, pp. 177–181, 2007.

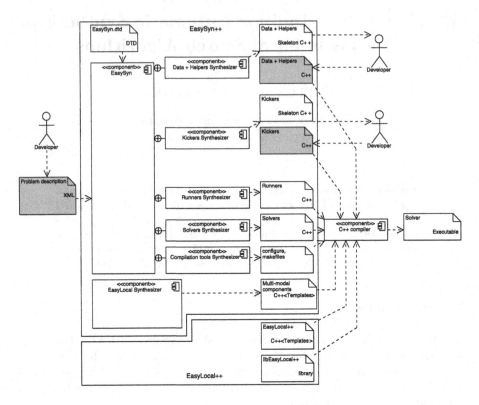

Fig. 1. The architecture of EASYSYN++

limited manual programming a first picture of the behavior of local search for his/her problem, although at the cost of a lot of computational power. However, being CPU time much cheaper than programmer's time, we believe that the use of such a tool could help in engineering local search algorithms.

The main difference with respect to other tools, such as for example Comet [5] and its predecessors, is that in those software environments the program corresponding to the specified algorithm is assembled internally by the system and therefore its execution is performed "behind the scenes". Conversely, EASY-SYN++ is completely *glass-box* (or *white-box*), and the exact C++ statements are "before user's eyes", and can be modified and adapted at user's preferences.

2   EasySyn++ Architecture

The architecture of EASYSYN++ is shown in Figure 1. The input is a XML description file (which refers to a DTD file for its validation) that is parsed by the core module, which calls the modules responsible for generating the code.

The description file specifies the local search structure and the set of runners, kickers and solvers that we want to synthesize. An example description file for the N-Queens problem is given in Listing 1.1. The `<data>` element specify the names of the classes and their role. In this case, for example, there is just one state representation (**<state>** element) and two neighborhoods (**<move>** elements).

The **<helpers>** element specifies the helpers, giving the data types they manage and the optional class names. Class names not supplied are generated by appending the standard name to the problem prefix (for example, NQ_StateManager).

The **<kickers>** and **<runners>** elements specify to which neighborhoods these components have to be applied. If more than one neighborhood is specified, Easy-Syn++ generates all possible ones that can be obtained using the union operator. In this context, the tag **<use-all-neighborhoods/>** declares that EasySyn++ has to synthesize one runner (or kicker) for each possible union neighborhood. For example, EasySyn++ generates three different HillClimbing runners: one for each move type and one for their union.

The number of generated runners is exponential in the number of neighborhoods: the size of the powerset minus one (the empty set). Therefore, when the number of neighborhoods is large, the use of **<use-all-neighborhoods/>** must be limited, and instead only subsets of the neighborhoods should be used.

Listing 1.1. The XML description file

```xml
<?xml version="1.0" standalone="no"?><!DOCTYPE problem SYSTEM "easysyn.dtd">
<problem name="nQueens" prefix="NQ">
<data dir="data/">
  <input classname="BoardSize" />
  <output classname="ChessBoard" />
  <state classname="QueensArray" />
  <move classname="Exchange" />
  <move classname="Swap" />
</data>
<helpers dir="helpers/">
  <state-manager state="QueensArray"/>
  <cost-component classname="Diagonal"/>
  <cost-component classname="Row"/>
  <neighborhood-explorer move="Exchange"/>
  <neighborhood-explorer move="Swap" classname="SwapExplorer"/>
  <output-manager output="ChessBoard"/>
  <prohibition-manager move="Exchange"/>
  <prohibition-manager move="Swap"/>
</helpers>
<kickers>
  <kicker id="sk"> <use-neighborhood ref="Swap"/> </kicker>
  <kicker id="ak"> <use-all-neighborhoods/> </kicker>
</kickers>
<runners>
  <runner type="HillClimbing" id="hc"> <use-all-neighborhoods/> </runner>
  <runner type="TabuSearch" id="ts"> <use-neighborhood ref="Exchange"/> </runner>
</runners>
<solvers>
  <solver type="IteratedLocalSearch"> <use-runner ref="ts"/> <use-kicker ref="ak" type="
      random"/> </solver>
  <solver type="TokenRing" max_length="3"> <use-all-runners/> </solver>
  <solver type="VariableNeighborhoodDescent"> <use-kicker ref="sk" type="best"/> </solver>
</solvers>
</problem>
```

Each module of EasySyn++ is responsible for a type of class to generate. The modules are briefly described in the following.

Data and helper synthesizer. This module creates the skeleton for Data and Helper classes. The actual code describing the attributes of states and moves of the application, together with the cost functions and the neighborhood exploration strategy, must be provided by the user.

Kickers Synthesizer. Kickers are synthesized almost completely. The human intervention regards only the notion of synergy. For each ordered pair of neighborhoods, EASYSYN++ defines a method of the kicker that takes a move of each one, and returns a boolean value. The body of the method must be supplied by the user by specifying the conditions on the values of the attributes of the moves to be synergic.

Runners Synthesizer. Runners are synthesized completely. This module has only the task to create the runner objects with the correct template instantiations and the links to the type-compatible helpers.

Solvers Synthesizer. Solvers also are synthesized completely. This module only creates the solver objects with the template instantiations and the links to runners and kickers. For each concrete solver there is a value of a command-line argument to invoke it. Therefore, all the necessary links with the runners are inserted one by one in a code branch guarded by the given value for such a command-line argument.

Compilation tools. This module does not generate C++ code but sets up a basic `configure` script and a set of `Makefiles` for building the system. The target compilation environment is the GNU *Autotools* suite, which comprises Autoconf, Automake and Libtool. This environment can be considered as the de-facto standard for portably building and installing applications across many systems (including various UNIX/Linux distributions, Mac OS X and Cygwin on Windows).

The result of the synthesis is a set of C++ source files and the portable compilation environment so that the generated code can be immediately compiled just out-of-the-box. Yet, the methods that must be supplied by the user are implemented as a single `throw` statement that raises a run-time exception when executed. This way the programmer is prompted to take care of the actual implementation of those methods, whereas if we had left the methods empty the programmer could forget to implement some method thus giving raise to errors whose effects could be unpredictable.

As shown in Figure 1, EASYSYN++ provides also the abstract EASYLO-CAL++ classes for *multimodal* runners and kickers based on neighborhood compositions, such as unions and sequences, as defined in [4]. These components are automatically synthesized for an arbitrary number k of neighborhoods, without the need to define new helper classes and to write any code.

EASYSYN++ provides also kickers for an arbitrary composition of neighborhoods (multimodal kickers). They provide methods for performing both random and best sequences of moves from the composite neighborhood.

In order to generate the best moves (restricted or general) the kicker uses a backtracking algorithm that backtracks as soon as two consecutive moves are not feasible.

3   Discussion and Conclusions

First, although important parts of the code must be provided by the user, the advantages of using EASYSYN++ are significant. Indeed, instead of writing the code from scratch, the user has only to insert specific fragments in the right places (indicated by a `throw` statement), without having to worry about object communication and high level control structures.

As a matter of fact, the programming effort needed to come up with a family of 16 solvers for the N-Queens problem requires only 168 user-supplied lines of code. The overall application is composed also of 750 synthesized lines of code, and 6,898 coming from EASYLOCAL++ (including the multimodal runners and kickers). Even though the number of lines is clearly a very rough measure of the human effort, this is very limited and EASYSYN++ can be considered as a software environment for the fast prototyping of SLS algorithms.

Another advantage of EASYSYN++ is that it makes an intensive and principled reuse of the code. For example, the exploration and the evaluation of the union of two or more neighborhoods is performed relying completely on the helpers of the underlying basic neighborhoods, without the need of any additional code. This design allows the user to avoid duplications that would make the maintenance of the code very problematic.

Needless to say, there are also some limits in using EASYSYN++ and EASY-LOCAL++. Their architecture prescribes a precise structure for the design of a local search algorithm: the entities of the problem are factorized in groups of related classes in the framework and the user is forced to use them in a controlled way. If, on the one hand, this helps in term of conceptual clarity and it is one of the main sources of software reuse, on the other hand, since the control logic is completely defined at the framework level, this feature poses some restrictions in the design of *ad hoc* SLS algorithms that need to be tightly tailored to some specific feature of the problem at hand.

References

1. Di Gaspero, L., Schaerf, A.: EasyLocal++: An object-oriented framework for flexible design of local search algorithms. Software—Practice and Experience 33(8), 733–765 (2003)
2. Di Gaspero, L., Schaerf, A.: Writing local search algorithms using EasyLocal++. In: Voß, S., Woodruff, D.L. (eds.) Optimization Software Class Libraries. OR/CS series, pp. 155–176. Kluwer Academic Publishers, Boston (2002)
3. Di Gaspero, L., Roli, A., Schaerf, A.: EasyAnalyzer: an object-oriented framework for the experimental analysis of stochastic local search algorithms. In: Stützle, T., Birattari, M., Hoos, H. (eds.) Engineering Stochastic Local Search Algorithms (SLS-2007). LNCS, vol. 4638, pp. 76–90. Springer, Heidelberg (2007)
4. Di Gaspero, L., Schaerf, A.: Neighborhood portfolio approach for local search applied to timetabling problems. Journal of Mathematical Modeling and Algorithms 5(1), 65–89 (2006)
5. Van Hentenryck, P., Michel, L.: Constraint-Based Local Search. MIT Press, Cambridge (MA), USA (2005)

Human-Guided Enhancement of a Stochastic Local Search: Visualization and Adjustment of 3D Pheromone

Jaya Sreevalsan-Nair[1], Meike Verhoeven[1], David L. Woodruff[1], Ingrid Hotz[2], and Bernd Hamann[1]

[1] University of California, Davis, CA, USA
[2] Konrad-Zuse-Zentrum für Informationstechnik, Berlin, Germany
{mverhoev,jnair,dlwoodruff,bhamann}@ucdavis.edu, hotz@zib.de

1 Introduction

In this paper, we describe user interaction with an optimization algorithm via a sophisticated visualization interface that we created for this purpose. Our primary interest is the tool itself. We demonstrate that a user wielding this tool can find ways to improve the performance of an ant colony optimization (ACO) algorithm as applied to a problem of finding 3D paths in the presence of impediments [14]. One part of a solution method can be to find a path on a grid. Of course, there are near linear time algorithms for the shortest path that have been applied to problems that are quite large. However, for a grid in three dimensions with arcs on the axes and diagonals, the problems can become extremely large as resolution is increased and heuristics thus make sense (see, e.g., [6] for state-of-the art algorithms where pre-processing is possible). Ant colony optimization (see, e.g., [4,5]) is ideally suited to such a problem.

The visualization literature covers many related aspects. For example, analysis of search algorithm results can be effectively done using visualization [7]. Monitoring an algorithm using visualization gives the user all levels of information [3]. Visualization plays a major role in steering computations as can be seen in numerous cases. A few instances are: visualization for interactive visual computing of programs [1], visualization in adaptive grid methods for debugging as well as analyzing computational algorithms [9]. Computational steering, which is the terminology for use of visualization to monitor simulations or iterations and using the feedback for improving computational results, has been a powerful research tool and a taxonomy of existing systems can be found in [11]. Mitsubishi research labs has conducted extensive research concerning what they call "human guided search" (see, e.g., [8,10]). They developed visualization tools for optimization problems such as scheduling, routing, and layout. One of their ideas is that computer algorithms can locate local minima given a starting point, but users can suggest good starting points. Our approach here continues this research line, but we provide tools for interacting directly with the operation of the ant system by allowing the user to lay down pheromone.

T. Stützle, M. Birattari, and H.H. Hoos (Eds.): SLS 2007, LNCS 4638, pp. 182–186, 2007.

2   Problem Formulation and Solution Method

We create a finite set of points $\mathcal{G} \subset \mathcal{A}$, then a path is defined by the variables p_{ij}, where p_{ij} is one if the arc from i to j is in used by the path and zero otherwise. The problem on a grid $\mathcal{G}$ is called (G) and is written as a shortest path problem between points a and b in $\mathcal{G}$:

$$\min_{p} \sum_{i \in \mathcal{G}} \sum_{j \in \mathcal{G}} \left[D_{ij} p_{ij} \left(1 + \sum_{s \in \mathcal{S}} c_s \beta(s, i, j) \right) \right] \quad \text{(G)}$$

$$\text{subject to: } \sum_{i \in \mathcal{G}} I_{ai} p_{ai} = 1; \quad \sum_{i \in \mathcal{G}} I_{ib} p_{ib} = 1$$

$$\sum_{i \in \mathcal{G}} I_{ik} p_{ik} - \sum_{j \in \mathcal{G}} I_{kj} p_{kj} = 0, \quad k \in \mathcal{G} \backslash \{a, b\}$$

where the elements of the incidence matrix, I_{ij}, have the value one if i and j are neighbors and zero otherwise. The elements of the distance matrix, D_{ij} give the distances between neighboring points i and j. The penalty for travel through impediment s in the set of spheres $\mathcal{S}$ is given by c_s. The data $\beta(s, i, j)$ gives the portion of the line segment between i and j that is in the range of impediment s (for large instances, this is computed only as needed). We use a rectangular grid with resolution k, with arcs on the coordinate axes as well as all of the diagonals.

In ant colony optimization, each ant decides independently which node to go to based on the pheromone trail τ_{ij} and a heuristic value η_{ij} for grid points i and j. We set our heuristic value so that it considers both path length and number of arcs:

$$\eta_{ij} = \begin{cases} ((\frac{\sqrt{3}}{k-1} + |i - b| - |j - b|) \cdot \gamma + \frac{1}{D_{ij}(1 + \sum_{s \in \mathcal{S}} c_s \beta(s,i,j))} \cdot (1 - \gamma), & \text{if } I_{ij} = 1 \\ 0, & \text{otherwise} \end{cases}$$

where γ is a parameter on which aspect to emphasize more. The first term computes the difference in distance from j and i to the end point, b. The maximum length of an arc is added to ensure that this term is always positive. The second term gives the cost for this arc.

The ant t located in the node i selects the arc to node j according to the probability

$$\frac{(\tau_{ij}(t))^{\alpha} (\eta_{ij})^{\beta}}{\sum_{n \in N} (\tau_{in}(t))^{\alpha} (\eta_{in})^{\beta}}, \quad \forall j \in N$$

where the parameters α and β determine the relative influence of pheromone and heuristic value. After each ant reaches the endpoint, pheromone is deposited:

$$\Delta_{ij}(t) = \begin{cases} Q/L(t) \text{ if } (i, j) \in T(t) \\ 0 \qquad \text{otherwise} \end{cases}$$

where Q is a constant and $L(t)$ is the length of the path $T(t)$ generated by ant t. So the shorter a path is the more pheromone is laid on its arcs. The amount of pheromone is updated according to the rule:

$$\tau_{ij}(t+1) = \rho\tau_{ij}(t) + \Delta\tau_{ij}(t)$$

where ρ ($0 < \rho < 1$) represents the persistence of the pheromone trail.

It is well known that some form of randomization is needed to improve the exploration done by the ants (see, e.g., [13,2]). The scheme given by Nakamichi and Arita [12] has been shown to be effective and it can be applied directly to our algorithm. We introduce a random selection rate r to improve the diversity of the paths by occasionally ignoring the heuristic selection function to select a random neighbor.

An important feature of our algorithm is the insertion of the user into the process, which is done by "correcting" an ant's path. On arcs that are deemed to be useful by the user extra pheromone is added. In order to do that in three dimensions, a powerful visualization tool is needed.

3   Visualization and Path Edit Tool

The primary functions of our visualization tool are viewing the current data and interactively manipulating the paths to help in faster convergence of our iterative algorithm in a 3D environment. The data to be visualized consists of paths, coordinates of the centers of spherical impediments, and their respective radii. These spheres are clipped by a bounding box, which is our volume of interest. Each sphere imposes a path-length penalty, which is indicated by the opacity of the sphere. As we generate paths using the ACO algorithm, they are plotted as thin solid tubes winding through the translucent spheres.

Each path is represented as a vector of points. An existing path can be modified by inserting a point, deleting a point, or overwriting a point. The tool facilitates various features to interactively modify the paths, and save these changes to feed it back into the algorithm. Hence our tool includes an interface for modifying an existing point or adding a new point. This point modifying tool shows the current coordinates of the point and allows the user to move along the three orthogonal axes to place the point in a new position.

The color of the path is determined by the cumulative (penalized) distance at each point on the path and is obtained by linearly interpolating a rainbow color scheme between zero and the maximum penalized distance possible. This allows us to have two different coloring possibilities- a global or a local coloring scheme for each path. The global scheme makes use of the extremum of the paths, while the local scheme uses only the corresponding penalty. (The penalties at the ends of the path are the extrema, as we are considering cumulative, penalized distance here.)

For ease of visualization, the tool also gives different options for viewing the spheres: using the penalty-based opacity, using user-defined opacity or as wireframe. It gives navigational control to the user to rotate and translate the bounding box and its contents. Since this tool is used for research purpose, the interaction with the ACO algorithm is via a very simple, file based interface: The graphics tool can be viewed and tested by visiting http://graphics.idav.ucdavis.edu/ jaya/InterdictionViz.

4 Experimental Results

We did some experiments with the limited goal of showing that we can use the tool to find algorithmic improvements. Using the visualization tool, we are able to find ways that could be used to refine the ACO. To show the improvement of path developed by ants when the user is in the loop we generated six different problem instances on a grid with resolution 50. These instances can be solved exactly, but our goal is to use visualization as an engineering tool. We found out that for these instances the ACO parameters $(\alpha, \beta, \gamma, \rho, Q, r, \delta) = (2, 3, 0.9, 0.99, 3, 0.03, 8)$ work well. Results are summarized in Table 1. All but the first and last columns show the best solution found after the given number of ants. The column labeled "1000with" shows the results if a hint is given after 500 ants. The first column gives the instance number and the last column gives the number of ants needed to find a solution without a hint that is as good as the solution found with the hint. So the user supplied adjustment is clearly quite valuable. This can be used to improve algorithms used for problem instances that are too large to be solved exactly.

Table 1. Results for the six instances

instance	500	1000with	1000	1500	2000	2500	ants until reached
1	2.880	2.180	2.546	2.283	2.224	2.170	2469
2	2.677	1.863	2.371	2.202	2.104	2.059	>500000
3	26.635	19.885	21.159	20.120	19.373	17.893	1643
4	8.544	5.935	7.514	7.113	6.908	6.854	11780
5	7.707	4.879	7.035	6.712	6.087	6.072	>500000
6	2.981	1.995	2.451	2.199	2.199	2.048	3545

5 Conclusions

We have described a powerful engineering tool for visualizing and interacting with an ACO, which is challenging in a 3D environment. Ant systems are particularly well suited for visualization, but our main contribution has been to demonstrate the use of a tool for interacting with the algorithm by allowing the user to lay down pheromone and hence influence the future paths taken. Computational experiments demonstrate the value of this approach to engineering stochastic local search algorithms. The tool has been shown to result in improved performance of the algorithm by allowing the user to take advantage of their knowledge of the problem and convey that via the search control parameters. Human aided optimization is a fairly young topic and seems to offer substantial promise as part of efforts to engineer stochastic local search algorithms.

References

1. Brunner, J.D., Jablonowski, D.J., Bliss, B., Haber, R.B.: Vase: the visualization and application steering environment. In: Supercomputing '93: Proceedings of the 1993 ACM/IEEE conference on Supercomputing, pp. 560–569. ACM Press, New York (1993)
2. Bullnheimer, B., Hartl, R., Strauss, C.: A new rank based version of the ant system. Central European Journal for Operations Research and Economics 7, 25–38 (1999)
3. Card, S.K., Mackinlay, J.D., Shneiderman, B. (eds.): Readings in information visualization: using vision to think. Morgan Kaufmann Publishers Inc. San Francisco, CA, USA (1999)
4. Dorigo, M., Gambardella, L.M.: Ant algorithms for discrete optimization. Artificial Life 5, 137–172 (1999)
5. Dorigo, M., Maniezzo, V., Colorni, A.: Ant system: optimization by a colony of cooperating agents. IEEE Transactions on Systems, Man, and Cybernetics - Part B 26, 1–13 (1997)
6. Goldberg, A.V., Kaplan, H., Werneck, R.: Reach for A*: Efficient point-to-point shortest path algorithms, `http://research.microsoft.com/research/pubs/view.aspx?type=Technical%20Report&id=986`
7. Hammond, S.P.: Putting the user in the loop: On-line user adaption of genetic algorithms. In: Sarker, R., Reynolds, R., Abbass, H., Chen Tan, K., McKay, B., Essam, D., Gedeon, T. (eds.) Proceedings of the 2003 Congress on Evolutionary Computation CEC2003, pp. 892–897. IEEE Press, Los Alamitos (2003)
8. Klau, G.W., Lesh, N.B., Marks, J.W., Mitzenmacher, M.: Human-guided tabu search. In: National Conference on Artificial Intelligence (AAAI), pp. 41–47 (2002)
9. Kreylos, O., Tesdall, A.M., Hamann, B., Hunter, J.K., Joy, K.I.: Interactive visualization and steering of CFD simulations. In: VISSYM'02: Proceedings of the Symposium on Data Visualisation 2002, Aire-la-Ville, Switzerland, pp. 25–34. Eurographics Association (2002)
10. Lesh, N., Lopes, L.B., Marks, J., Mitzenmacher, M., Schafer, G.T.: Human-guided search for jobshop scheduling. In: 3rd International NASA Workshop on Planning and Scheduling for Space (October 2002)
11. Mulder, J.D., van Wijk, J.J., van Liere, R.: A survey of computational steering environments. Future Gener. Comput. Syst. 15(1), 119–129 (1999)
12. Nakamichi, Y., Arita, T.: Diversity control in ant colony optimization. Artificial Life and Robotics 7, 1614–7456 (2004)
13. Stützle, T., Hoos, H.: $\mathcal{MAX} - \mathcal{MIN}$ ant system. Future Generations Computer Systems Journal, 16 (2000)
14. Verhoeven, M., Woodruff, D.L.: Optimizing paths in the presence of spherical impediments. Technical report, GSM, UC Davis, Davis CA 95616 (2007)

Solving a Bi-objective Vehicle Routing Problem by Pareto-Ant Colony Optimization

Joseph M. Pasia[1], Karl F. Doerner[2,3], Richard F. Hartl[2], and Marc Reimann[4]

[1] Department of Mathematics, University of the Philippines-Diliman,
Quezon City, Philippines
[2] Department of Business Administration, University of Vienna, Vienna, Austria
[3] Salzburg Research Forschungsgesellschaft, Salzburg, Austria
[4] Institute for Operations Research, ETH Zurich, Zurich, Switzerland
jmpasia@up.edu.ph, {karl.doerner, richard.hartl}@univie.ac.at,
marc.reimann@ifor.math.ethz.ch

Abstract. In this paper we propose the application of Pareto ant colony optimization (PACO) in solving a bi-objective capacitated vehicle routing problem with route balancing (CVRPRB). The objectives of the problem are minimization of the tour length and balancing the routes. We propose PACO as our response to the deficiency of the Pareto-based local search (P-LS) approach, which we also developed to solve CVR-PRB. The deficiency of P-LS is the lack of information flow among its pools of solutions. PACO is a natural choice in addressing this deficiency since PACO and P-LS are similar in structure. It resolves the absence of information flow through its pheromone values. Several test instances are used to demonstrate the contribution and importance of information flow among the pools of solutions. Computational results show that PACO improves P-LS in most instances with respect to different performance metrics.

1 Introduction

The vehicle routing problem (VRP) [1] consists of finding the optimal route for a fleet of vehicles, starting and ending at a single depot, that must serve a set of n customer demands such that each customer is visited by only one vehicle route. If each vehicle can only collect a maximum capacity of Q units of demand, then the problem is known as capacitated VRP (CVRP).

Although exact approaches have been proposed to solve the VRP, the last 15 years witnessed increasing research effort on the development of metaheuristic approaches since the VRP has been proven $\mathcal{N}$P-hard [2]. Likewise, metaheuristics have also been used to solve vehicle routing problems with $k \geq 2$ objectives.

One example of a multiobjective VRP is the CVRP with route balancing (CVRPRB), where the two objectives of the CVRPRB are (i) to minimize the sum of the total distance travelled by each vehicle and (ii) to minimize the difference between the longest and shortest vehicle tours. The CVRPRB was tackled using population-based local search (P-LS) [3]. This algorithm repeatedly

T. Stützle, M. Birattari, and H.H. Hoos (Eds.): SLS 2007, LNCS 4638, pp. 187–191, 2007.

generates a pool of good initial solutions by using a randomized savings algorithm followed by local search.

In this paper, we propose a new approach in solving the CVRPRB by improvin P-LS. The decision to improve P-LS is based on our own observation that, although it has been shown that this algorithm outperformed other state of the art metaheuristics, it does not have any mechanism that allows the pools of solutions to share information. Furthermore, there is also no sharing of information among individuals or solutions of each pool. We believe that by addressing these deficiencies, better results may be generated.

To allow cooperation among individuals and pools, we use Pareto Ant Colony Optimization (PACO), which was first developed to solve a multiobjective portfolio optimization problem [4]. This metaheuristic was an application of Ant Colony Optimization (ACO) [5] to multiobjective problems.

In our proposed approach, the individuals of each pool share information via the local pheromone update of PACO. This update allows the individuals to explore the regions of the search space that are not yet visited by the previous individuals. On the other hand, pools share information indirectly by the local pheromone update and directly by the global pheromone update. The global pheromone update allows the current pool to lead the next pool towards the better region.

2   Pareto Ant Colony Optimization for CVRPRB

2.1   Initialization Phase of PACO

Just like P-LS, the starting solutions of PACO are initialized by a pool S of identical solutions. These identical solutions assign each customer to a separate tour. These solutions are then improved by combining customers i and j using the pheromone information values τ_{ij}^k. It is worth mentioning that customers i and j must belong to the candidate list created by the savings algorithm [6]. This strategy of creating the list was first introduced in [7].

It follows that given the set Ω_γ of γ feasible combinations having the largest savings, the combination $(i, j) \in \Omega_\gamma$ is chosen with probability

$$(i,j) = \begin{cases} \arg\max_{(i,j)\in\Omega_\gamma} \sum_{k=1}^{2} w_i^k \cdot \tau_{ij}^k & \text{if } q \le q_0 \\ \widehat{(i,j)} & \text{otherwise} , \end{cases} \tag{1}$$

where the random variable $\widehat{(i,j)}$ has a probability distribution given by

$$\Pr_{ij} = \begin{cases} \dfrac{\sum_{k=1}^{2} w_i^k \cdot \tau_{ij}^k}{\sum_{(u,l)\in\Omega_\gamma} \left(\sum_{k=1}^{2} w_u^k \cdot \tau_{ul}^k\right)} & (i,j) \in \Omega_\gamma \\ 0 & \text{otherwise.} \end{cases} \tag{2}$$

After a solution has been improved, the 2-exchange operator is applied immediately in order to avoid artificially balanced solutions [8].

Local pheromone update. After applying the 2-exchange operator, the individual then updates the pheromone values using the equation $\tau_{ij}^k = (1 - \rho) \cdot \tau_{ij}^k + \rho \cdot \tau_{\min} \; \forall k$, where (i, j) is an edge belonging to the individual, ρ is the evaporation rate (with $0 \leq \rho \leq 1$) and $\tau_{\min}$ is a small number that serves as the minimum possible pheromone value. This update evaporates the pheromone values only along the edges visited by an individual in order to allow the succeeding individuals to explore other edges.

2.2 Local Search Phase of PACO

Both PACO and P-LS consist of three local search operators namely, move, swap, and 2-exchange. The move operator inserts a customer of one vehicle tour to another vehicle tour. The swap operator interchanges two customers of two vehicle tours. The 2-exchange operator is then called every time the move or swap operators are applied.

Starting from a solution $z \in \mathcal{S}$, the local search phase explores the feasible candidate solutions of the move neighborhood $\mathcal{N}_1$ of z. These candidate solutions are evaluated using dominance relation i.e., all feasible neighboring solutions are compared and the dominated ones are removed. Each of the remaining non-dominated solutions will undergo the same process as z i.e., its entire neighborhood is searched and all the dominated solutions are removed. We repeat the entire process of searching the whole neighborhood and removing the dominated solutions until all solutions in the neighborhood are dominated. When this happens, we perform the local search process on the set $\mathcal{P}$ of all non-dominated solutions found so far using the swap neighborhood $\mathcal{N}_2$. Unless the termination conditions are not satisfied after performing $\mathcal{N}_2$, we apply $\mathcal{N}_1$ again on the Pareto set returned by $\mathcal{N}_2$. Since exploring the entire neighborhood of a given solution is computationally expensive, only a certain number of solutions are allowed to undergo local search [3].

Global pheromone update. After the local search phase, the global pheromone update is performed. This update intensifies the pheromone values along the edges that are found in the best and second best solutions of each generation t. In other words, the best and second best solutions w.r.t. each objective update the pheromone values in order to guide the initialization phase at generation $t + 1$. The global pheromone update is given by the equation $\tau_{ij}^k = (1 - \rho) \cdot \tau_{ij}^k + \rho \cdot \triangle \tau_{ij}^k$ where (i, j) is an edge in the first and/or second best solution, and the quantity $\triangle \tau_{ij}^k$ is 15 if (i, j) is an edge in the best and 2nd best solutions of objective k, 10 if (i, j) is an edge only in the best solution of objective k, and 5 (i, j) is an edge only in the 2nd best solution of objective k.

3 Numerical Results

The numerical analysis was performed on a set of benchmarks described in [9]. The set of benchmarks consists of 7 test instances having 50 to 199 customers and a single depot. Three of the instances were generated such that the customers are uniformly distributed on a map and the remaining instances feature clusters of customer locations. All test instances have capacity constraints. We run all our methods on a personal computer with a 3.2 GHz processor; the algorithms were coded in C++ and compiled using the GCC 4.1.0 compiler.

3.1   Evaluation Metrics and Parameter Settings

Unary quality indicators have become standard tools in assessing the performance of different algorithms for bi-objective optimization problems. They complement the traditional approach of using graphical visualization, which may provide information on how the algorithm works [10]. This study considered three unary quality indicators namely, the *hypervolume indicator* [11], the *unary epsilon indicator*, and the *R3 indicator* [12].

The parameters of PACO that are also parameters of P-LS were left unchanged from their original values as used in [3]. For instance, the initial number of solutions in each pool S is equal to the number of customers and the size of the candidate list is given by $\lfloor 0.10 \times (\# \text{ of customers}+1) \rfloor$. The values of ρ, q_0, and τ_{min} are 0.05, 0.75, and 0.00001 respectively.

3.2   Analysis

Ten runs with different random seeds were performed for each of the test instances. Before applying the different unary indicators, all approximation sets are normalized between 1 and 2. The reference set for each test instance consists of the points that are not dominated by any of the approximation sets generated by all algorithms under consideration.

Based on the values of the different unary quality indicators, we observed the following: First, for instances with 50 and 75 customers (small-size instances), PACO does not seem to improve the unary quality indicators of P-LS. Second, for instances with 100 and 120 customers (medium-size instances), one can clearly see that the hypervolume values of PACO are much better than that of P-LS. The corresponding unary epsilon and R3 indicator values of PACO are slightly better than that of P-LS. PACO also recorded the best values for the three unary quality indicators. Third, PACO greatly improves the quality of solutions of P-LS in instances with 150 and 199 customers (large-size instances). Their unary quality indicators are much superior to that of P-LS.

From these findings, one may note that as the test intances grow in size, the performance of PACO becomes much better than P-LS. This may indicate that sharing information among populations and among solutions is beneficial for medium and large instances.

4   Conclusion

In this paper, we improved the Pareto-based local search (P-LS) recently developed to solve a bi-objective vehicle routing problem. Our improvement is based on our observation that P-LS did not have any mechanism for allowing its population to share information. For improving P-LS, we used Pareto-ant colony optimization (PACO). This metaheuristic allows the exchange of information among the populations and among the solutions through the pheromone values. In particular, PACO carries information during the local and global pheromone updates. PACO is a natural choice for improving P-LS since both have similar algorithmic structure. Computational results showed that PACO improved the performance of P-LS with respect to the unary quality indicators, as the size of the test instance grows bigger.

References

1. Dantzig, G., Ramsey, J.: The truck dispatching problem. Management Science 6, 80–91 (1959)
2. Lenstra, J., Kan, A.: Complexity of vehicle routing and scheduling problem. Networks 11, 221–227 (1981)
3. Pasia, J.M., Doerner, K.F., F., H.R., Reimann, M.: A population-based local search for solving a bi-objective vehicle routing problem. In: Cotta, C., van Hemert, J. (eds.) Evolutionary Computation in Combinatorial Optimisation - EvoCOP 2007. LNCS, vol. 4446, pp. 166–175. Springer, Heidelberg (2007)
4. Doerner, K., Gutjahr, W., Hartl, R., Strauss, C., Stummer, C.: Pareto ant colony optimization: A metaheuristic approach to multiobjective portfolio selection. Annals of Operations Research 131, 79–99 (2004)
5. Dorigo, M., Maniezzo, V., Colorni, A.: Ant system: Optimization by a colony of cooperating agents. IEEE Transactions on Systems, Man, and Cybernetics - Part B 26(1), 29–41 (1996)
6. Clarke, G., Wright, J.: Scheduling of vehicles from a central depot to a number of delivery points. Operations Research 12, 568–581 (1964)
7. Doerner, K., Gronalt, M., Hartl, R.F., Reimann, M., Strauss, C., Stummer, M.: SavingsAnts for the vehicle routing problem. In: Cagnoni, S., Gottlieb, J., Hart, E., Middendorf, M., Raidl, G.R. (eds.) EvoIASP 2002, EvoWorkshops 2002, EvoSTIM 2002, EvoCOP 2002, and EvoPlan 2002. LNCS, vol. 2279, pp. 11–20. Springer, Heidelberg (2002)
8. Jozefowiez, N., Semet, F., Talbi, E.: Parallel and hybrid models for multi-objective optimization: Application to the vehicle routing problem. In: Guervós, J.J.M., Adamidis, P.A., Beyer, H.-G., Fernández-Villacañas, J.-L., Schwefel, H.-P. (eds.) Parallel Problem Solving from Nature - PPSN VII. LNCS, vol. 2439, pp. 271–280. Springer, Heidelberg (2002)
9. Christofides, N., Mingozzi, A., Toth, P.: The vehicle routing problem. In: Christofides, N., Mingozzi, A., Toth, P., Sandi, C. (eds.) Combinatorial Optimization, John Wiley and Sons, Chichester (1979)
10. Knowles, J., Thiele, L., Zitzler, E.: A tutorial on the performance assessment of stochastic multiobjective optimizers. Technical Report TIK-Report No. 214, Computer Engineering and Networks Laboratory, ETH Zurich, Switzerland (2006)
11. Zitzler, E., Thiele, L.: Multiobjective evolutionary algorithms: a comparative case study and the strength Pareto approach. IEEE Trans. Evolutionary Computation 3(4), 257–271 (1999)
12. Hansen, M., Jaszkiewicz, A.: Evaluating the quality of approximations to the non-dominated set. Technical Report Technical Report IMM-REP-1998-7, Technical University of Denmark (1998)

A Set Covering Approach
for the Pickup and Delivery Problem
with General Constraints on Each Route

Hideki Hashimoto[1], Youichi Ezaki[2], Mutsunori Yagiura[3], Koji Nonobe[4],
Toshihide Ibaraki[5], and Arne Løkketangen[6]

[1] Graduate School of Informatics, Kyoto University, Kyoto, Japan
[2] Canon System Solutions Inc., Japan
[3] Graduate School of Information Science, Nagoya University, Nagoya, Japan
[4] Faculty of Engineering, Hosei University, Koganei, Japan
[5] School of Science and Technology, Kwansei Gakuin University, Sanda, Japan
[6] Molde University College, Molde, Norway
hasimoto@amp.i.kyoto-u.ac.jp, yezaki@amp.i.kyoto-u.ac.jp,
yagiura@nagoya-u.jp, nonobe@k.hosei.ac.jp, ibaraki@ksc.kwansei.ac.jp,
Arne.Lokketangen@himolde.no

Abstract. We consider a generalization of the pickup and delivery problem with time windows by allowing general constraints on each route, and propose a heuristic algorithm based on the set covering approach, in which all requests are required to be covered by a set of feasible routes. Our algorithm first generates a set of feasible routes, and repeats reconstructing of the set by using information from a Lagrangian relaxation of the set covering problem corresponding to the set. The algorithm then solves the resulting set covering problem instance to find a good feasible solution for the original problem. We conduct computational experiments for instances with various constraints and confirm the flexibility and robustness of our algorithm.

1 Introduction

In the pickup and delivery problem with time windows (PDPTW) we are given a set of requests, where each request signifies the delivery of a load from an origin to a destination [1,2]. The origin and destination of each request must be visited by the same vehicle in the order of origin and destination. Each service (i.e., pickup at an origin or delivery at a destination) must start within a given time window. Each vehicle has a capacity and the total amount of loads of a vehicle cannot exceed its capacity.

In this paper, we consider a generalization of the pickup and delivery problem with time windows by allowing general constraints on each route (abbreviated as PDP-GCER). We assume that the constraints on each route satisfy the monotone property: If a route consisting of a set of requests satisfies a constraint, then any subroute (i.e., consisting of a subset of the requests visited in the same order) also satisfies the constraint. We allow any monotone constraint provided that we can

T. Stützle, M. Birattari, and H.H. Hoos (Eds.): SLS 2007, LNCS 4638, pp. 192–196, 2007.
© Springer-Verlag Berlin Heidelberg 2007

determine its feasibility in a reasonable time. We also assume the traveling times satisfy the triangle inequality, which implies that time window constraints satisfy the monotone property. Note that many constraints that appear in practical situations are often monotone.

In our algorithm, we formulate the problem as the set covering problem (abbreviated as SCP), in which all requests must be covered by a set of feasible routes. Since enumerating all feasible routes is not realistic for reasonable problem sizes, the algorithm constructs a set of good routes which is of manageable size but has sufficient diversity. It constructs the initial set of routes by an insertion method, and then repeats a reconstruction procedure. The reconstruction procedure estimates the attractiveness of each route by the relative cost of the Lagrangian relaxation of the set covering problem with the current set of routes and generates new routes from those with small relative costs by applying five types of operations. The algorithm then solves the resulting set covering problem instance to find a good feasible solution of PDP-GCER. This type of approach, called column generation, tends to be efficient for problems with complicated or tight constraints. Note that our algorithm is a heuristic algorithm though the column generation method is usually used for exact algorithms.

2   Problem Definition

Let $G = (V, E)$ be a complete directed graph with vertex set $V = \{0, 1, \ldots, 2n\}$ and edge set $E = \{(i, j) \mid i, j \in V, i \neq j\}$. In this graph, vertex 0 is the depot and other vertices are customers where a load is picked up or delivered. Each edge $(i, j) \in E$ has a traveling cost $c_{ij} \geq 0$ and a traveling time $t_{ij} \geq 0$ and they satisfy the triangle inequalities. Let $H = \{1, 2, \ldots, n\}$ be the given set of requests. Each request $h \in H$ signifies the delivery from the origin $h \in V$ to the destination $h + n \in V$ (for convenience, we call a request and its origin by the same name h). The vertices h and $h+n$ must be visited by the same vehicle, and h must be visited before $h + n$. We consider the problem of serving all requests by a fleet of homogeneous vehicles. Each vehicle must start from the depot and return to the depot. Let S_r be the set of requests served in route r, $m_r = |S_r|$, and σ_r be the sequence of customers to be visited, where $\sigma_r(k)$ denotes the kth customer in r. We assume $\sigma_r(0) = \sigma_r(2m_r + 1) = 0$.

In this paper, we consider various constraints on each route. Each customer $i \in V$ has a handling time s_i for the service and a time window $[e_i, l_i]$, where e_i is the release time to serve i and l_i is the deadline of the service. Each request h consumes q_{hp}^{re} units of renewable resources ($p = 1, 2, \ldots, \rho$) while it is loaded, and consumes $q_{hp'}^{\mathrm{non}}$ units of nonrenewable resources ($p' = 1, 2, \ldots, \pi$). Each vehicle has capacities Q_p^{re} for renewable resource p and $Q_{p'}^{\mathrm{non}}$ for nonrenewable resource p'. The total load of each renewable resource p at each customer in route r must not exceed the capacity Q_p^{re}. The total load of each nonrenewable resource p' must be within $Q_{p'}^{\mathrm{non}}$. We further introduce the Last-In First-Out (abbreviated as LIFO) constraint. That is, if a request h is picked up before a request h', either h is delivered before the pickup of h' or after the delivery of h'. Note that

LIFO constraint satisfies the monotone property. As for the LIFO constraint, we consider both cases where the constraint is imposed and not. In addition to the above constraints, any monotone constraint can be introduced, assuming that we have an algorithm to test its feasibility efficiently.

Let ν be the number of vehicles used in a solution. A feasible solution is a set $\{\sigma_1, \sigma_2, \ldots, \sigma_\nu\}$ of routes such that each σ_r satisfies all the given constraints and each request is serviced exactly once. The objective function is $\sum_{r=1}^{\nu} C_r$, where $C_r = \alpha + \sum_{i=0}^{2m_r} c_{\sigma_r(i)\sigma_r(i+1)}$ (i.e., C_r is the sum of a fixed cost α for using a vehicle and the traveling cost of r).

3   Set Covering Approach

The PDP-GCER can be formulated as the set covering problem SCP(R^*): $\min\{\sum_{r \in R^*} C_r x_r \mid \sum_{r \in R^*} a_{hr} x_r \geq 1, \forall h \in H, x_r \in \{0,1\}, \forall r \in R^*\}$ where R^* is the set of all feasible routes, and $a_{hr} = 1$ if request h is in route $r \in R^*$, otherwise $a_{hr} = 0$.

Since enumerating all feasible routes is not realistic, we choose a subset R of manageable size from all feasible routes R^* and solve the corresponding set covering problem SCP(R) by an SCP solver. Finally we obtain a solution of PDP-GCER from the solution of SCP(R). The solution to SCP(R) may contain more than one route serving the same requests. In this case, we can remove the over-covered requests one by one in a greedy way until no such request remains. The main part of our algorithm is how to determine the set R. To obtain a good solution, R should be chosen very carefully. The generation of a set of routes consists of two phases. The first phase is the initial construction phase, which generates a certain number of routes for each request by an insertion method. The second phase is the reconstruction phase, which chooses good routes from the current set of routes, and generates their neighboring routes. The algorithm executes the initial construction phase once, and then repeats the reconstruction phase until a given time limit is reached.

The initial construction phase starts from the empty set $R = \emptyset$, and generates a certain number of routes for each request by an insertion method. The insertion method first prepares a route that contains a single request and the depot, and then repeats inserting requests into the route. If it always inserts a request which achieves the minimum insertion cost, the resulting set of routes may not have sufficient diversity. We therefore incorporate randomness in the algorithm based on the idea often used in GRASP. When the route becomes maximal (i.e., no more request can be inserted to it), we add it to R.

In the reconstruction phase, the algorithm first calculates the Lagrangian multipliers by a subgradient method. The Lagrangian relaxation problem of SCP(R) for a given nonnegative Lagrangian multiplier vector $\boldsymbol{u} = (u_1, u_2, \ldots, u_n)$ is defined as $L(\boldsymbol{u}) = \min\{\sum_{r \in R} c_r(\boldsymbol{u}) x_r + \sum_{h \in H} u_h \mid x_r \in \{0,1\}, \forall r \in R\}$ where $c_r(\boldsymbol{u}) = C_r - \sum_{h \in H} a_{hr} u_h$ is the relative cost associated with r. It is known that $L(\boldsymbol{u})$ gives a lower bound on the optimal value of problem SCP(R). If a good Lagrangian multiplier vector $\boldsymbol{u}$ is obtained, the relative cost $c_r(\boldsymbol{u})$ gives reliable information on the attractiveness of fixing $x_r = 1$ because any r with $x_r = 1$ in an optimal solution of SCP tends to have a small $c_r(\boldsymbol{u})$ value. The algorithm chooses routes with small relative costs for each request in order to generate new routes from them by applying the following five types of operations. We introduce three methods to generate neighboring routes of a route r. An insertion operation inserts a request h into r at the best position. This operation is applied for each request (which is not in r). A deletion operation deletes one request from r. This operation is applied for each request in r. A swap operation deletes one request from r and then inserts one request which is not in r at the best position. This operation is applied for all pairs of a request in r and another not in r. All feasible routes obtained by these operations are generated. In addition, the algorithm uses two operations to generate neighboring routes of two routes r and r'. A 2-opt* operation is applied to two routes r and r' that satisfy $S_r \cap S_{r'} = \emptyset$. This operation first constructs a route by concatenating a former part of r and a latter part of r' at positions k and k' respectively. As the resulting route may not satisfy all constraints, it then modifies the route by insertion and/or deletion operations to satisfy the violated constraints. For given two routes r and r', a mixing operation starts from $\sigma_{\mathrm{mix}} := \sigma_r$ and repeats modifying the current route σ_{mix} so that the set of requests in it becomes closer to that of $\sigma_{r'}$ by inserting or deleting different requests between the two routes σ_{mix} and $\sigma_{r'}$. All routes obtained during the modifications are considered as candidates to be added into R.

4   Computational Experiment

The algorithm was coded in C and run on a PC (Intel Pentium4, 2.8 GHz, 1 GB memory). We compared our algorithm with a metaheuristic algorithm coded in reference to the algorithm proposed for PDPTW by Li and Lim [3]. It is based on the simulated annealing and tabu search procedure based on the same objective function as ours. We modify it so that it can deal with PDP-GCER. The modified algorithm executes the local search in the feasible region under the constraints of PDP-GCER.

We generated the PDP-GCER instances consisting of six groups GC1–GC6 from a part of Li and Lim's PDPTW instances [3] by adding various constraints to them where each group contains 18 instances. In Table 1, columns "ρ" and "π" represent the number of renewable and nonrenewable resources. Column "TW" shows the information about the time window constraint. In GC4 and GC5, we set all the time windows (i.e., that of all customers and the depot) to $[0, \infty)$. On the other hand, in GC3, we shortened the original time windows by 4%. For the rest (i.e., GC1, GC2 and GC6), we adopted the time windows of the original instances. We imposed the LIFO constraint to GC2 and GC5 as shown in the LIFO column by 1. The time limit of constructing routes is set to 2400 seconds and the time limit of solving the set covering problem is set to 1200 seconds for our algorithm. We set the time limit to 3600 seconds for the metaheuristic algorithm.

Table 1. Comparison for GC1–GC6

	Resource				Ours		LS	
INST	ρ	π	TW	LIFO	CNV	CDIST	CNV	CDIST
GC1	1	0	–	0	208	65624.54	224	72422.65
GC2	3	1	–	1	278	95016.41	313	92170.04
GC3	1	0	−4%	0	142	48421.68	155	56234.36
GC4	1	1	$[0, \infty)$	0	234	79763.98	212	59545.98
GC5	1	1	$[0, \infty)$	1	238	84378.57	212	55065.95
GC6	2	0	–	0	271	84785.49	276	82716.75

Column "LS" represents the results of metaheuristic algorithm, column "CNV" means the cumulative number of vehicles and column "CDIST" means the cumulative traveling cost. The results show that for GC1, GC2, GC3 and GC6 whose instances have multiple constraints or severe constraints, our algorithm works efficiently, but for GC4 and GC5 whose instances have weaker constraints, the metaheuristic algorithm works better than ours.

5   Conclusion

We generalized the pickup and delivery problem with time windows by allowing general constraints, and propose an algorithm that generates a set of feasible routes and applies the set covering approach. The algorithm constructs an initial set of routes by an insertion method and reconstructs the set of routes repeatedly by modifying the routes using various types of neighborhood operations while reducing the candidate routes by utilizing the Lagrangian relative costs. The computational results indicated that our algorithm works more efficiently than a metaheuristic algorithm for instances having tighter constraints.

References

1. Bent, R., Hentenryck, P.V.: A two-stage hybrid algorithm for pickup and delivery vehicle routing problems with time windows. Computers and Operations Research 33, 875–893 (2006)
2. Ropke, S., Pisinger, D.: An adaptive large neighborhood search heuristic for the pickup and delivery problem with time windows. Transportation science 40(4), 455–472 (2006)
3. Li, H., Lim, A.: A metaheuristic for the pickup and delivery problem with time windows. International Journal on Artificial Intelligence Tools 12(2), 173–186 (2003)

A Study of Neighborhood Structures for the Multiple Depot Vehicle Scheduling Problem[*]

Benoît Laurent[1,2] and Jin-Kao Hao[2]

[1] Perinfo SA, Strasbourg, France
[2] LERIA, Université d'Angers, Angers, France
blaurent@perinfo.com, hao@info.univ-angers.fr

Abstract. This paper introduces the "block moves" neighborhood for the Multiple Depot Vehicle Scheduling Problem. Experimental studies are carried out on a set of benchmark instances to assess the quality of the proposed neighborhood and to compare it with two existing neighborhoods using shift and swap. The "block moves" neighborhood can be beneficial for any local search algorithm.

1 Introduction

Given $\mathcal{T} = \{t_1, t_2, \ldots, t_n\}$ a set of trips, a fleet of vehicles housed in $\mathcal{K} = \{n + 1, n + 2, \ldots, n + m\}$ depots, each having a limited capacity r_k, the Multiple Depot Vehicle Scheduling Problem (MDVSP) consists of determining a least-cost feasible vehicle schedule. Each trip t_i is defined by an origin and a destination with associated starting and ending times (s_i, e_i). We denote by τ_{ij}, the travel time from the end location of trip t_i to the starting location of trip t_j. Two trips are said compatible if t_j can be achieved right after t_i by the same vehicle, i.e. $e_i + \tau_{ij} \leq s_j$. Transfers without passengers are called deadhead trips. Transfers either to come from or return to the depot are the pull-out and pull-in trips respectively. The set of vehicles can also be defined by $\mathcal{V} = \{1, 2, \ldots, p\}$. The fleet is supposed to be homogeneous such that trips can be performed by any vehicle. In a valid schedule, the trips performed by each vehicle are pairwise compatible. The vehicles start from and return to their depot. Finally, the number of vehicles at each depot does not exceed the depot capacity.

The main objective of the MDVSP is to minimize the number of vehicles in use. Other objectives aim to avoid non-commercial tasks that induce traveling and/or waiting costs. They are denoted by c_{ij} for deadhead trips connecting t_i and t_j, by c_{ki} for the pull-out trip starting at depot d to reach trip t_i and by c_{jk} for the pull-in trip from t_j to return to the depot d. Note that fixed vehicle costs are added to pull-in and pull-out trips costs.

The MDVSP is NP-hard when two depots at least are considered [1]. The literature offers a panel of solution methods. Early works focused on heuristic algorithms. Exact algorithms have been proposed since the end of the 1980s (for

---

[*] Partially supported by the French Research Ministry (CIFRE No. 176/2004).

T. Stützle, M. Birattari, and H.H. Hoos (Eds.): SLS 2007, LNCS 4638, pp. 197–201, 2007.
© Springer-Verlag Berlin Heidelberg 2007

a thorough survey, see [2] and [3]. In [4], Large Neighborhood Search (LNS) and Tabu Search (TS) were employed for the first time on the MDVSP.

The main contribution of this paper is the introduction and an in-depth study of a new neighborhood schema for the MDVSP. This new neighborhood, called "block moves", is based on the notion of ejection chains. We report comparative studies of the two neighborhoods used in the TS algorithm of [4] and of our "block moves" neighborhood.

2   Solution Approach

2.1   Decision Variables, Domains and Constraints

The set of decision variables is the set of trips T, the domain $\mathcal{D}_i$ associated to each variable t_i corresponding to vehicles. A configuration is an assignment of vehicles in $\mathcal{V}$ to trips in T. It can be represented as a vector of integers:

$$\sigma = (\sigma(t_1), \sigma(t_2), \ldots, \sigma(t_n)) \qquad (\sigma \in \mathcal{V}^n)$$

The search space Ω is then defined as the set of all such assignments.

To obtain a feasible configuration σ, the compatibility constraint between trips must be satisfied: $\forall\, (t_i, t_j) \in T^2,\ \sigma(t_i) = \sigma(t_j),\ e_i + \tau_{ij} \le s_j$

2.2   Evaluation Function

The evaluation function measures the quality of a solution σ. In addition to the costs previously described, it comprises a penalty term for constraint violations.

$$\forall \sigma \in \Omega, \quad f(\sigma) = w_c f_c(\sigma) + f_o(\sigma) \tag{1}$$

where $w_c > 0$ is the weight associated to the constraints violations, f_c the number of violations detected in σ, f_o the value of the objective function on σ.

2.3   Initial State

The initial solution is built by means of a greedy algorithm, relying on a Forward Checking procedure. At each step, the choice of variables follows the min-domain heuristic, ties being broken randomly. Each selected variable (a trip) is labeled by a possible vehicle with a priority given to already employed vehicles.

3   Neighborhood Structures

3.1   Existing Neighborhood Structures

We first describe the neighborhoods, $\mathcal{N}_{shift}$ and $\mathcal{N}_{swap}$, embedded in the Tabu Search of [4]. In $\mathcal{N}_{shift}$, a neighbor is obtained by transferring a trip to another vehicle. The swap move implies two trips t_i and t_j, accomplished by two different vehicles v and v'. It consists in moving t_i from v to v' and t_j from v' to v. The respective size of $\mathcal{N}_{shift}$ and $\mathcal{N}_{swap}$ is $O(np)$ and $O(n^2)$. Each time a change concerns the first or the last trip of a vehicle, we check if a transfer of the entire sequence of trips to an available vehicle of another depot would be profitable.

3.2   Block-Moves Neighborhood

The block-moves neighborhood is a parameterized structure, based on ejection chains. The principle of ejection chains was introduced in [5].

Let bl be the size of the "block", that is the number ($bl \geq 1$) of trips that will initiate the sequence of moves. Our neighborhood mechanism, called $\mathcal{N}_{bl\_moves}$, consists in moving bl consecutive trips handled by the same vehicle v to another vehicle v'. These ejection moves often cause constraints violations that will trigger repair attempts. For each conflicting trip, we scan and retain the best vehicle, that may receive it. If no such vehicle exists, the conflicts remain. As in $\mathcal{N}_{shift}$ and $\mathcal{N}_{swap}$, complete transfers of trips to vehicles belonging to other depots are evaluated. The size of $\mathcal{N}_{bl\_moves}(\sigma)$ depends on the number of vehicles running at least bl trips in the current configuration σ.

The rationale behind $\mathcal{N}_{bl\_moves}$ is the following. First, it prevents the search from being stuck in local optima because of some conflicts that could be repaired as soon as they arise. Second, behind the notion of "block moves", we aim at preserving the good properties of the configuration, typically the trips that fit well together.

4   Computational experiments

4.1   Benchmarks and Experimental Settings

Our experiments rely on the benchmarks proposed in [4] and generated as in [6]. In these instances, $n \in \{500, 1000, 1500\}$, $m \in \{4, 8\}$ and the cost incurred for the use of a vehicle is 10000. We heavily penalized constraints violations (100000 each) to discourage the exploration of infeasible regions.

$\mathcal{N}_{bl\_moves}$ is by definition parameterized by bl. In our experiments, bl ranges from 1 to 5. Enlarging this domain was considered useless since the average number of trips per vehicle is approximately equal to 4.2.

4.2   Statistical Study of the Neighborhoods

To assess the performance differences between $\mathcal{N}_{shift}$, $\mathcal{N}_{swap}$ and $\mathcal{N}_{bl\_moves}$, we computed some statistics on the neighborhood of the initial solutions issued as described in section 2. Table 1 gathers the average values (over 20 independent runs) on the instances m4n1000s0 to m4n1000s4, arbitrary chosen since all results were similar.

The second column displays the size of the neighborhoods, the two next ones the percentages of improving (Imp.) and deteriorating moves (Det.) respectively. The percentage of plateau moves, almost negligible, is omitted. The five next columns focus on the improving moves. They contain the values of the average improvement (Avg), the standard deviation (Sd), the best improvement (Best), the Hamming distance (d_H) between σ and σ' and the associated Standard deviation (Sd). The average value and the distance to the best known solution are indicated in the table caption (respectively, Avg value and Gap). These

Table 1. m4n1000 (heuristic) - Avg. Value = 2574683.67 - Gap (%) = 3.88

	Size	Imp. (%)	Det. (%)	Improvements				
				Avg	Sd	Best	d_H	Sd
shift	429000	0.41	99.34	167.50	(156.24)	1036.22	2.71	(2.52)
swap	497582	0.17	99.83	186.27	(177.59)	1161.66	3.67	(2.34)
1_moves	427052	6.64	93.30	275.09	(245.77)	2046.83	4.95	(3.04)
2_moves	323991	5.01	94.93	326.66	(289.90)	2144.54	7.15	(3.22)
3_moves	204345	4.66	95.02	321.09	(304.14)	2158.02	8.12	(3.55)
4_moves	113503	5.07	94.48	296.62	(302.21)	2094.69	8.42	(3.67)
5_moves	58289	5.54	94.16	299.03	(304.72)	2018.69	9.21	(3.73)

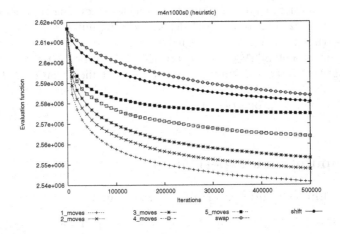

Fig. 1. Evolution of the cost function according to the neighborhood mechanism

percentages must not be compared to those of [4] in which solutions values have been purged of vehicles costs. Here, all costs are taken into account.

We observe that $\mathcal{N}_{bl\_moves}$ clearly outperforms the other neighborhoods independently of the value of bl. The probability of obtaining an improving neighbor in $\mathcal{N}_{bl\_moves}$ is much higher than in $\mathcal{N}_{shift}$ or $\mathcal{N}_{swap}$. Moreover, the average and best improvements procured by $\mathcal{N}_{bl\_moves}$ are about twice superior to the values related to the shift and swap neighborhoods.

To further investigate the neighborhoods behavior, we observe the profile of the evaluation function during the search of a *first improvement Descent*[1] algorithm, using the competing neighborhoods. The stop criterion is based on the number of iterations elapsed since the last improvement. This number was set to the size of the neighborhood. Figure 1 shows, for the instance m4n1000s0, the mean evolution (on 20 runs) of the evaluation function (y-axis) during a search process (x-axis). One observes that the $\mathcal{N}_{bl\_moves}$ neighborhood always lead to a better convergence with respect to the $\mathcal{N}_{shift}$ or $\mathcal{N}_{swap}$ neighborhoods.

Concerning the influence of the bl parameter, Figure 1 shows that it might be not worthwhile to consider blocks of size strictly greater than 3 for these

---

[1] A best improvement descent would be too time-comsuming.

instances. This value is certainly relative to the average number of trips per vehicle (approximately 4.2).

Finally, let us indicate making use of a portfolio of 3 simple *Descent* algorithms based on the *bl_moves* neighborhood ($bl \in \{1, 2, 3\}$), cyclically applied in a diversification scheme (see [7]), we obtained results of good quality. For example, on the instances containing 4 depots and 500 trips, the average gap from optimality is of 6.195 again 10.919 for TS [4].

5   Conclusion

In this paper, we investigated for the first time in the context of the MDVSP, a fundamental component of any local search algorithm, namely the neighborhood structure. We designed a new parameterized neighborhood schema, called "block-moves" and compared it with existing neighborhoods.

The computational study carried out on a set of artificial instances clearly shows the advantage of our "block-moves" neighborhood. A portfolio of the most effective neighborhood structures is being exploited to tackle the MDVSP with promising results. We are convinced that integrating the "block-moves" neighborhood in other frameworks can be beneficial.

References

1. Bertossi, A., Carraresi, P., Gallo, G.: On some matching problems arising in vehicle scheduling models. Networks 17, 271–281 (1987)
2. Odoni, A.R., Rousseau, J.M., Wilson, N.H.: Models in urban and air transportation, vol. 6, pp. 107–150. Elsevier Science, North-Holland, Amsterdam (1994)
3. Desaulniers, G., Hickman, M.: Public transit, pp. 69–120. Elsevier Science, North-Holland, Amsterdam (2007)
4. Pepin, A.S., Desaulniers, G., Hertz, A., Huisman, D.: Comparison of heuristic approaches for the multiple vehicle scheduling problem. Technical Report EI2006-34, Economie Institute, Erasmus University Rotterdam, Rotterdam (2006)
5. Glover, F.: Ejection chains, reference structures and alternating path methods for traveling salesman problems. Discrete Applied Mathematics 65(1-3), 223–253 (1996)
6. Carpenato, G., Dell'Amico, M., Fischetti, M., Toth, P.: A branch and bound algorihm for the multiple depot vehicle scheduling problem. Networks 19, 531–548 (1989)
7. Di Gaspero, L., Schaerf, A.: Neighborhood portfolio approach for local search applied to timetabling problems. Journal of Mathematical Modeling and Algorithms 5(1), 65–89 (2006)

Local Search in Complex Scheduling Problems

Thijs Urlings and Rubén Ruiz

Instituto Tecnológico de Informática, Universidad Politécnica de Valencia,
Valencia, Spain
thijs_urlings@iti.upv.es, rruiz@eio.upv.es

Abstract. In this paper different local search procedures are applied to a genetic algorithm for a complex hybrid flexible flow line problem. General precedence constraints among jobs are taken into account, as are machine release dates, time lags and sequence dependent setup times; both anticipatory and non-anticipatory. Although closely connected to real-world problems, this combination of constraints is hardly treated in the literature. This paper presents a study of the behaviour of local search for such a complex problem. A combination of the local search variants is presented. Comprehensive statistical experiments indicate that substantial improvements in solution quality can be reached.

1 Introduction

Local search is a widely applied and praised solution improvement technique. Many heuristics and meta-heuristics in combinatorial optimization employ such techniques. Some researchers have tried to fill the gap between the operations research literature related to scheduling, where relatively easy problems are solved with high level algorithms, and the production literature, where myopic algorithms such as dispatching rules are implemented for more realistic problems. In these cases, some realistic constraints are considered but still the problems approached are either not general enough or specific to some production scheduling scenarios. Additionally, the issue of how local search operators behave in such complex problems appears to be unaddressed.

This paper addresses a heavily constrained hybrid flexible flow line (HFFL) problem. General precedence constraints among jobs are taken into account. Release dates are given for the machines, as are time lags between stages. This problem also considers sequence dependent setup times that can be either anticipatory or non-anticipatory. This combination of constraints implies a close connection to real-world industrial problems. For the considered problem, various local search neighbourhoods are compared. A combination of the neighbourhood searches within a genetic algorithm is also tested. The algorithm called SGA, presented and evaluated in Urlings et al. [3], is used as a basis. The solution representation consists of one job sequence for all stages and a machine assignment rule.

T. Stützle, M. Birattari, and H.H. Hoos (Eds.): SLS 2007, LNCS 4638, pp. 202–206, 2007.

2   Problem Description

The hybrid flexible flow line problem can be described as follows (see also Ruiz et al. [1]): Given is a set of jobs $N = \{1, \ldots, n\}$ to be processed on a production line, consisting of a set of stages $M = \{1, \ldots, m\}$. Each stage i, $i \in M$ contains a set of unrelated machines $M_i = \{1, \ldots, m_i\}$.

The flexibility of the problem implies that jobs might skip stages. Each job $j \in N$ visits a set of stages $F_j \subseteq M$. The processing time for job j on machine l at stage i is denoted p_{ilj}. We consider the following constraints: $E_{ij} \subseteq M_i$ is the set of eligible machines for job j in stage i; $P_j \subset N$ gives set of predecessors of job j; rm_{il} expresses the release date for machine l in stage i; lag_{ilj} models the time lag for job j between stage i and the next stage to be visited, when job j is processed on machine l at stage i; S_{iljk} denotes the setup time between the processing of job j and job k on machine l inside stage i and A_{iljk} is a binary parameter that indicates whether the corresponding setup is anticipatory or not. The objective is to minimize the largest completion time, known in literature as makespan. As reasoned in Ruiz et al. [1], the problem is $\mathcal{NP}$-Hard.

3   Local Search Variants

Probably the most important decision in local search design is the definition of the neighbourhood. Large neighbourhoods are powerful, but time consuming. The smallest neighbourhood we implement is Adjacent Interchange (AI), which consists in interchanging pairs of adjacent jobs. The pair of adjacent jobs whose interchange causes the largest decrease of the makespan, is interchanged. Because of precedence relationship, the number of neighbours is less or equal $n - 1$.

To increase the neighbourhood size one can check, besides the adjacent position, also the position at a distance of two (DoT). In our implementation, if the job at position j goes to position $j + 2$, both the job at position j and at $j + 1$ are shifted one position to the front. Of course precedence relationships have to be taken into account again. Therefore, a maximum of $(n - 1) + (n - 2) = 2n - 3$ neighbours are considered at each neighbourhood evaluation.

A more extensive search is allowed if all positions for a job are evaluated (APJ). For precedence constraints, the number of reinsertion positions is generally lower than $n - 1$; the job can only be inserted after its last predecessor and before its first successor. The permutation with the best makespan value is used as the new permutation.

4   Accelerations

As many similar permutations have to be compared, it seems straightforward to implement some accelerations. The faster a local search is, the more searches can be done (or generations made by the GA) per unit of time. However, the complexity of the problem we consider limits the possibilities to accelerate. The accelerations by Taillard [2], for example, are not applicable.

What can be done is using the part of the permutation that is unaffected by the movements. Suppose that the jobs in the positions j and $j + 1$ are interchanged. Then, the tasks of the jobs until position $j - 1$ remain unaffected (for $j > 1$).

Note that this only holds for non look-ahead machine assignment rules. Furthermore, precedence constraints among jobs complicate the matter entirely. Forgetting about precedence constraints for the moment, the number of job calculations without accelerations for AI is $(n - 1)n$. With accelerations, we need to calculate the complete schedule for the first two neighbours, where respectively the first two and the second and the third job are interchanged. The third neighbour has the same job in the first position as the second neighbour and can use its completion times. The total number of job calculations is consequently $n + n + (n - 1) + \ldots + 3 = n(n + 1)/2 + n - 3$. For large n this gets close to 50%. For the other neighbourhoods, comparable results are to be found. The precedence constraints, however, reduce the time advantage. Suppose that, in an example with n jobs (n divisible by 3 without loss of generality), only the job at position $n/3$ and at position $n/2$ can be interchanged with their neighbours. Then $n + (1 + 2n/3) = 5n/3 + 1$ job calculations have to be made, which is a time advantage of only about $1/6$ compared to the regular $2n$ calculations. The advantage in practice is studied next.

5   Experimental Evaluation

For our experiments we use a subset of the benchmark provided by Ruiz et al. [1], publicly available at `www.upv.es/gio/rruiz/` with the best known solution values. The subset contains the 192 large instances with 50 or 100 jobs.

All experiments are executed on a Pentium IV computer with a single 3.0 GHz processor and 1 GB of RAM memory. The algorithm is implemented in Delphi with the 2006 compiler and run under Windows XP Professional.

5.1   Comparison of Local Search Techniques

To study the influence of the accelerations and the effectiveness of the local search techniques, we run 800 generations of the SGA and then apply local search to the final population. Local search is repeated until the solutions are not improved anymore. APJ is applied to all jobs in random order. For the 800 generations, computation times range from 30 seconds for the smaller, till 15 minutes for the largest instances. We carry out tests with and without accelerations.

In Table 1 we show the results. The impact of the accelerations depends on the size of the neighbourhood. For APJ, an average time saving of about 28.5% is measured. The DoT local search with accelerations is about 40.1% faster than the version without accelerations. For AI this is even 46.8%.

The displayed makespan deviation is the average relative deviation from the best known solution value. Relative improvement is the relative difference between this last deviation and the average deviation without local search. As expected, smaller neighbourhood sizes have less possibilities to improve solutions, but have a large advantage in computation time.

Table 1. Average Relative Percentage Deviation in time and makespan for the local search variants

LS variant	Time dev. orig.	Time dev. acc.	Makespan dev.	Rel. improvement
no LS	-	-	4.863	-
AI	12.35	6.57	4.622	4.948
DoT	28.48	17.07	4.544	6.543
APJ	1002.97	717.62	3.917	19.439

5.2   A Compound Local Search Method

In the previous section we have applied local search to the final population as a exploratory measure to observe the potential of the technique. In this section we are interested in embedding the local search in the genetic algorithm itself. Each time after creating two new individuals in the SGA algorithm, local search is applied with probability pLS. The procedure is not applied to the (possibly poor) new individuals, but to an already accepted individual in the population.

AI-search is applied to one of the individuals with the best makespan. If no improvement is made, in the next iteration AI-search is applied to one of the individuals with the second-best makespan value. When all makespans have had one individual AI-investigated, the investigated individual with lowest makespan is taken for DoT-search. Once all makespan values have had their individual DoT-searched, APJ is applied on a random job for the same individuals in the same order. When all jobs have been investigated, we know that the individuals are local optima for all implemented LS neighbourhoods and we start investigating the remaining individuals of the population, with makespans equal to the locally optimal individuals.

The number of individuals in the population with the same makespan plays an important role in this procedure. A new individual is only accepted if it is better than the worst individual in the population and if the permutation does not exist in the population yet and if the number of individuals with this same makespan does not exceed a given number $max\#sol$. Preliminary tests show that applying LS only after half of the allowed CPU time is far more efficient than applying LS directly from the start. It seems plausible to allow the SGA to carry out the initial coarse search which is in turn also faster than with LS.

To test the configuration, we compare pLS equal to 0, 10% and 100% and $max\#sol$ equal to 1, 15 and 200 (total population). The algorithm is executed five times for each combination. We define the allowed running time as $25 \cdot n \cdot \sum_i m_i$ milliseconds. In Figure 1 the interactions and the 99% Tukey confidence intervals are shown. The best results are obtained for $pLS =10\%$. With respect to 100%, a smaller part of the running time is consumed and the genetic algorithm has more power. A maximum number of 15 individuals with the same makespan is optimal. Although some of the confidence intervals overlap, using local search decreases the relative makespan deviation a 5.0% from 4.00% to 3.80%, using the same CPU time. The improvement compared to the original algorithm is 10.9%, which is in turn a significant improvement. As a preliminary conclusion,

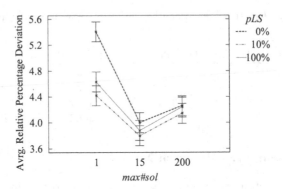

Fig. 1. Interaction between local search probability and $max\#sol$

the more elaborated local search has improved the results substantially when compared under the same stopping CPU time.

6   Conclusions

In this paper, we have evaluated various local search neighbourhood implementations for a highly constrained hybrid flexible flowshop problem. We have shown the limited possibilities of applying regular accelerations and the efficiency of these. The consequences of each neighbourhood search are shown.

A new compound way of applying local search within an algorithm is presented and an important adaptation is made to the original method. Stopping on passed CPU time, however, results in a 5.0% solution quality improvement when local search is applied. The complexity of the problem is a clear disadvantage for local search techniques. However, due to a change in the acceptance criterion, a significant total improvement of 10.9% is made compared to the original algorithm. Contrary to simple scheduling problems, where straightforward local search techniques are frequently applied, it seems that such techniques are neither easy nor apparently as profitable when applied to much more complex environments. As a result, more focused research efforts are needed in this direction.

References

1. Ruiz, R., Sivrikaya Şerifoğlu, F., Urlings, T.: Modeling realistic hybrid flexible flowshop scheduling problems. Computers and Operations Research (in press)
2. Taillard, E.: Some efficient heuristic methods for the flow shop sequencing problem. European Journal of Operational Research 47, 65–74 (1990)
3. Urlings, T., Ruiz, R., Sivrikaya Şerifoğlu, F.: Genetic algorithms for complex hybrid flexible flow line problems. Departamento de Estadística e Investigación Operativa Aplicadas y Calidad. Universidad Politécnica de Valencia, Technical Report DEIOAC-2007-02 (March 2007)

A Multi-sphere Scheme for
2D and 3D Packing Problems*

Takashi Imamichi and Hiroshi Nagamochi

Department of Applied Mathematics and Physics, Graduate School of Informatics,
Kyoto University, Kyoto, Japan
{ima,nag}@amp.i.kyoto-u.ac.jp

Abstract. In this paper, we deal with a packing problem that asks to
place a given set of objects such as non-convex polytopes compactly
in $\mathbb{R}^2$ and $\mathbb{R}^3$, where we treat translation, rotation and deformation as
possible motions of each object. We propose a multi-sphere scheme that
approximates each object with a set of spheres to find a compact layout
of the original objects. We focus on the case that all objects are rigid, and
develop an efficient local search algorithm based on a nonlinear program
formulation.

Keywords: packing problem, multi-sphere scheme, iterated local search,
unconstrained nonlinear program.

1   Introduction

In this paper, we deal with a packing problem that asks to place a given set of
objects such as non-convex polytopes or objects with curved surfaces compactly
in a bounded space in $\mathbb{R}^2$ and $\mathbb{R}^3$, where we treat translation, rotation and
deformation as possible motions of each object. The problem includes various
types of packing problems such as the circle/sphere packing problem, the non-
convex polygon packing problem and a problem of finding feasible layouts in
protein docking. We propose a general approach, called *multi-sphere scheme*,
for designing efficient algorithms that compute compact layouts in the packing
problem. In the multi-sphere scheme, we first approximate each object by a set of
spheres, and we then search for the positions of all these spheres that minimize an
appropriate penalty function. Geometric computations such as collision detection
and penetration depth [1] become significantly simpler if all objects are sets of
spheres.

In this paper, we focus on the *penalized rigid sphere set packing problem* that
asks to pack a family of sphere sets in our multi-sphere scheme. The container is
a rectangle or a circle for $\mathbb{R}^2$ and a cuboid or a sphere for $\mathbb{R}^3$, and only translation
and rotation by an arbitrary angle are allowed as motions of each object.

---

* This research was supported by Research Fellowships of the Japan Society for the
Promotion of Science for Young Scientists and a Scientific Grant in Aid from the
Ministry of Education, Science, Sports and Culture of Japan.

T. Stützle, M. Birattari, and H.H. Hoos (Eds.): SLS 2007, LNCS 4638, pp. 207–211, 2007.
© Springer-Verlag Berlin Heidelberg 2007

Numerous types of packing problems have been studied so far. The problem of packing circles/spheres, which can be regarded as a special case of the penalized rigid sphere set packing problem, has many variations, such as packing circles into a circle [2], packing circles into a rectangle [3], and sphere packing problem in $\mathbb{R}^2$ or a higher dimensional space [4]. Wang et al. [2] formulated an unconstrained nonlinear program of the problem, and designed an algorithm by combining a steepest descent method and a procedure for perturbing a layout.

The irregular strip packing problem has been well studied recently. The problem asks to place polygons in $\mathbb{R}^2$ inside a rectangular container so that the length of the container is minimized, where translations and rotations by fixed degrees such as 90 degrees and 180 degrees are usually allowed. Imamichi et al. [5] proposed a *separation* algorithm based on a nonlinear program and incorporated the separation algorithm and a swapping algorithm of two polygons in their iterated local search algorithm. Milenkovic [6] uses a linear program to compute how to translate and rotate polygons in a given layout in $\mathbb{R}^2$ so as to reduce the length of the container.

Modeling an object by a sphere set is already used in several applications. Ferrez [7] proposed a framework of physical simulation of a set of spheres in $\mathbb{R}^3$ and a fast algorithm to detect the collisions of spheres.

To design a local search algorithm for penalized rigid sphere set packing problem, we formulate the problem as an unconstrained nonlinear program such that the quasi-Newton method can be applied. We then construct an iterated local search algorithm based on the approach by Imamichi et al. [5].

2   Penalized Rigid Sphere Set Packing Problem

This section formulates the penalized rigid sphere set packing problem for $\mathbb{R}^3$, which asks to place a collection $\mathcal{O} = \{O_1, \ldots, O_m\}$ of m objects into a container C, which is a sphere or a cuboid. Each object O_i consists of n_i spheres $\{S_{i1}, \ldots, S_{in_i}\}$, and has a reference point r_i. Let c_{ij} be the vector that represents the center of sphere S_{ij}, r_{ij} be the radius of S_{ij} ($i = 1, \ldots, m; j = 1, \ldots, n_i$), and $N = \sum_{i=1}^m n_i$. We let $r_i = \sum_{j=1}^{n_i} c_{ij}/n_i$, which represents the center of O_i. For a set of points S, let $\overline{S}$ be the complement of S, $\mathrm{cl}(S)$ be the closure of S, ∂S be the boundary of S, and $\mathrm{int}(S) = S \setminus \partial S$ be the interior of S.

After translating object O by translation vector $(x, y, z)^\mathsf{T}$, the resulting object is described as $O \oplus (x, y, z)^\mathsf{T} = \{x + (x, y, z)^\mathsf{T} \mid x \in O\}$, where $^\mathsf{T}$ denote the transpose of a vector/matrix. After rotating object O by z-x-z Euler angles (ϕ, θ, ψ), the resulting object is described as $R_3(\phi, \theta, \psi) O$, where $R_3(\phi, \theta, \psi)$ represents a rotation matrix. We define a function $\Lambda_{\alpha_{\mathrm{rot}}}$ that gives the placement of object O_i by $\Lambda_{\alpha_{\mathrm{rot}}}(O_i, v_i) = (R_3(\alpha_{\mathrm{rot}}\phi_i, \alpha_{\mathrm{rot}}\theta_i, \alpha_{\mathrm{rot}}\psi_i)(O_i \oplus (-r_i))) \oplus ((x_i, y_i, z_i)^\mathsf{T} + r_i)$, where α_{rot} is a positive parameter that represents sensitivity of rotations. Let $S_{ij}(v_i) = \Lambda_{\alpha_{\mathrm{rot}}}(S_{ij}, v_i)$ and $c_{ij}(v_i)$ be the center of $S_{ij}(v_i)$. The penetration depth [1] of two objects O and O' is defined by $\delta(O, O') = \min\{\|x\| \mid \mathrm{int}(O) \cap (O' \oplus x) = \emptyset, x \in \mathbb{R}^3\}$, where $\|\cdot\|$ denotes the Euclidean norm.

We define penalties of the penetration and protrusion by using the penetration depth: The penalized rigid sphere set packing problem for $\mathbb{R}^3$ is defined by

$$\text{minimize} \quad F_{\text{rigid}}(v) = w_{\text{pen}}F_{\text{pen}}(v) + w_{\text{pro}}F_{\text{pro}}(v),$$
$$\text{subject to} \quad v = (v_1, \ldots, v_m) \in \mathbb{R}^{6m}, \tag{1}$$

where $F_{\text{pen}}(v) = \sum_{1 \leq i < k \leq m} \sum_{j=1}^{n_i} \sum_{l=1}^{n_k} f_{ijkl}^{\text{pen}}(v)$, $F_{\text{pro}}(v) = \sum_{i=1}^{m} \sum_{j=1}^{n_i} f_{ij}^{\text{pro}}(v)$, $f_{ijkl}^{\text{pen}}(v) = (\delta(S_{ij}(v_i), S_{kl}(v_k)))^2$, $f_{ij}^{\text{pro}}(v) = (\delta(S_{ij}(v_i), \text{cl}(\overline{C})))^2$. f_{ijkl}^{pen} denotes the penetration penalty of two spheres S_{ij} and S_{kl}, f_{ij}^{pro} denotes the protrusion penalty of sphere S_{ij}, and w_{pen} and w_{pro} are positive parameters. The penalized rigid sphere set packing problem for $\mathbb{R}^2$ can be formulated analogously. The computation of (1) takes $O(N^2)$ by a naive algorithm; the computation of F_{pen} is the bottleneck.

2.1   Computation of the Gradient of the Objective Function

This subsection shows how to compute $\nabla F_{\text{pen}}(v)$. First, $\partial f_{ijkl}^{\text{pen}}(v)/\partial v_a = 0$ ($a \in \{1, \ldots, m\} \setminus \{i, k\}$). Assume that a pair of spheres $S_{ij}(v_i)$ and $S_{kl}(v_k)$ intersect each other. Let $(x_{ij}, y_{ij}, z_{ij})^{\mathsf{T}} = c_{ij}(v_i)$. If $c_{ij}(v_i) \neq c_{kl}(v_k)$,

$$\frac{\partial f_{ijkl}^{\text{pen}}(v)}{\partial x_i} = 2\beta_{\text{pen}}(x_{ij} - x_{kl}),$$

$$\frac{\partial f_{ijkl}^{\text{pen}}(v)}{\partial \phi_i} = 2\beta_{\text{pen}}(c_{ij}(v_i) - c_{kl}(v_k))^{\mathsf{T}} \cdot \frac{\partial R_3(\alpha_{\text{rot}}\phi_i, \alpha_{\text{rot}}\theta_i, \alpha_{\text{rot}}\psi_i)}{\partial \phi_i} \cdot (c_{ij} - r_i).$$

where $\beta_{\text{pen}} = -\delta(S_{ij}(v_i), S_{kl}(v_k))/\|c_{ij}(v_i) - c_{kl}(v_k)\|$. The other differentials are calculated in the same way. In the case of $c_{ij}(v_i) = c_{kl}(v_k)$, we use subgradients. We can handle $\nabla F_{\text{pro}}(v)$ analogously. Thus we can compute $\nabla F_{\text{rigid}}(v)$.

3   Algorithm

This section describes our iterated local search algorithm ILS_RIGID. Given a layout $(\mathcal{O}, C)$ as an initial solution to the penalized rigid sphere set packing problem (1), our local search algorithm RIGIDQN$(\mathcal{O}, C)$ returns a locally optimal solution computed by applying the quasi-Newton method. RIGIDQN has three parameters w_{pen}, w_{pro}, and α_{rot}: w_{pen} (w_{pro}) denotes the weights of the penetration penalty (resp. the protrusion penalty), and α_{rot} denotes the sensitivity of rotations. We let $w_{\text{pen}} = w_{\text{pro}} = 1$ and $\alpha_{\text{rot}} = 10^{-2}$ in our experiments.

ILS_RIGID$(\mathcal{O}, C)$ generates an initial solution by placing objects $\mathcal{O}$ randomly so that the reference points are in the container C, rotates the objects randomly and applies RIGIDQN to the initial layout. ILS_RIGID maintains the incumbent solution $\mathcal{O}_{\text{opt}}$ that minimizes the objective function F_{rigid} of (1), which will be used for generating the next initial solution. ILS_RIGID first perturbs the incumbent solution by swapping two randomly chosen objects O_1 and O_2 and rotates them randomly. Then, ILS_RIGID translates and rotates objects by invoking RIGIDQN. ILS_RIGID repeats these operations until a time limit.

4 Computational Results

This section reports the results on computational experiments. We implemented our algorithm ILS_RIGID in C++, compiled it by GCC 4.0.2 and conducted computational experiments on a PC with an Intel Xeon 2.8GHz processor (Net-Burst) and 1GB memory. We adopt a quasi-Newton method package L-BFGS [8] for algorithm RIGIDQN.

Table 1. Information of instances

2D						3D				
Instance	TNO	TNS	ANS	Width	Height		Instance	TNO	TNS	ANS	EL
dighe1	16	1415	88.43	100	110		mol1	2	2575	1287.5	54
shapes0	43	3176	73.86	40	60		mol2	3	1948	649.3	46
swim	48	5598	116.62	5752	6500		mol3	3	2679	893.0	51

TNO: The total number of objects   ANS: The average number of spheres
TNS: The total number of spheres   EL: The edge length

Table 2. Results

Instance	Avg. EL	#runs	Time limit	Avg. F_{rigid}	Best F_{rigid}	Avg. NL	Best NL
dighe1	105	10	600	2.72e+01	2.29e+00	4.97e-02	1.44e-02
shapes0	50	10	600	8.60e-01	4.48e-02	1.86e-02	4.23e-05
swim	6126	10	600	2.92e+02	0.00e+00	2.79e-03	0.00e+00
mol1	54	10	600	3.66e+00	1.14e+00	3.54e-02	1.97e-02
mol2	46	10	600	1.08e+01	1.50e+00	7.16e-02	2.67e-02
mol3	51	10	600	9.64e+00	3.66e+00	6.09e-02	3.75e-02

We use instances for $\mathbb{R}^2$ and $\mathbb{R}^3$. For $\mathbb{R}^2$ we consider the irregular strip packing problem with free rotations. We use instances available at the EURO Special Interest Group on Cutting and Packing (ESICUP) website[1] after converting each polygon in the instances into a set of circles that approximates the polygon. For $\mathbb{R}^3$, we consider the problem that asks to place some protein molecules into a cubic container. We use molecule data of a protein–protein docking benchmark set[2] [9]. To generate instances, we place spheres with van der Waals radii. We generate three instances: mol1 consists of 1GRN_l_u and 1F34_l_b, mol2 consists of 1AY7_l_b, 1EWY_l_b and 1FC2_r_u, and mol3 consists of 1MLC_l_u, 1KAC_l_b and 1AY7_l_b. Table 1 shows the information of the instances.

We conducted 10 runs under the time limit of 600 seconds per run. Since the size of containers varies, we normalize the values of F_{rigid} by $\sqrt{F_{rigid}}/$(the average edge length of the container). Table 2 shows the results. The column "Avg. EL" denotes the average edge length of the container, the column "Avg. NL" denotes the average normalized value of F_{rigid} over the 10 runs, and the column "Best NL" denotes the best normalized value of F_{rigid} over the 10 runs.

---

[1] http://www.fe.up.pt/esicup/
[2] http://zlab.bu.edu/julianm/benchmark/

5 Conclusions

We proposed a multi-sphere scheme for a packing problem in 2D and 3D. It approximates objects by sphere sets and find a feasible layout of the objects by the iterated local search algorithm ILS_RIGID. Our computational experiments indicate that ILS_RIGID can compute a nearby feasible layout efficiently.

It is left as future work to accelerate ILS_RIGID by adopting other sophisticated data structures such as those for the collision detection and to develop an algorithm for the packing problem with deformable objects.

References

1. Agarwal, P.K., Guibas, L.J., Har-Peled, S., Rabinovitch, A., Sharir, M.: Penetration depth of two convex polytopes in 3D. Nordic Journal of Computing 7(3), 227–240 (2000)
2. Wang, H., Huang, W., Zhang, Q., Xu, D.: An improved algorithm for the packing of unequal circles within a larger containing circle. European Journal of Operational Research 141(2), 440–453 (2002)
3. Birgin, E.G., Martínez, J.M., Ronconi, D.P.: Optimizing the packing of cylinders into a rectangular container: A nonlinear approach. European Journal of Operational Research 160(1), 19–33 (2005)
4. Sloane, N.J.A.: The sphere packing problem. Documenta Mathematica ICM III, 387–396 (1998)
5. Imamichi, T., Yagiura, M., Nagamochi, H.: An iterated local search algorithm based on nonlinear programming for the irregular strip packing problem. Technical report 2007-009, Department of Applied Mathematics and Physics, Graduate School of Informatics, Kyoto University (2007)
6. Milenkovic, V.J.: Rotational polygon overlap minimization and compaction. Computational Geometry 10(4), 305–318 (1998)
7. Ferrez, J.A.: Dynamic triangulations for efficient 3D simulation of granular materials. PhD thesis, École Polytechnique Fédérale de Lausanne (2001)
8. Liu, D.C., Nocedal, J.: On the limited memory BFGS method for large scale optimization. Mathematical Programming 45(3), 503–528 (1989)
9. Mintseris, J., Wiehe, K., Pierce, B., Anderson, R., Chen, R., Janin, J., Weng, Z.: Protein-protein docking benchmark 2.0: an update. Proteins 60(2), 214–216 (2005)

Formulation Space Search for Circle Packing Problems

Nenad Mladenović[1], Frank Plastria[2], and Dragan Urošević[3]

[1] School of Mathematics, Brunel University, West London, UK
[2] Vrije Universiteit Brussel, Brussel, Belgium
[3] Mathematical Instiute SANU, Belgrade, Serbia
Nenad.Mladenovic@brunel.ac.uk, Frank.Plastria@vub.ac.be,
draganu@mi.sanu.ac.yu

Abstract. Circle packing problems were recently solved via reformulation descent (RD) by switching between a cartesian and a polar formulation. Mixed formulations, with circle parameters individually formulated in either coordinate system, lead to local search methods in a formulation space. Computational results with up to 100 circles are included.

1 Introduction

Traditional ways to tackle an optimization problem consider a given formulation $\min\{f(x)|x \in \mathcal{S}\}$ and search in some way through its feasible set $\mathcal{S}$. The consideration that a same problem may often be formulated in different ways allows to extend search paradigms to include jumps from one formulation to another. Each formulation should lend itself to some traditional search method, its 'local search' that works totally within this formulation, and yields a final solution when started from some initial solution. Any solution found in one formulation should easily be translatable to its equivalent formulation in any other formulation. We may then move from one formulation to another using the solution resulting from the former's local search as initial solution for the latter's local search. Such a strategy will of course only be useful in case local searches in different formulations behave differently.

This idea was recently investigated in [6] using an approach that systematically changes formulations for solving circle packing problems (CPP). There it is shown that a stationary point of a non-linear programming formulation of CPP in Cartesian coordinates is not necessarily stationary so in a polar coordinate system. The method *Reformulation descent* (RD) that alternates between these two formulations until the final solution is stationary with respect to both is suggested. Results obtained were comparable with the best known values, but they were achieved some 150 times faster than by an alternative single formulation approach. In that same paper we also introduced the idea suggested above of *Formulation space search* (FSS), using more than two formulations. Some research in that direction has been reported in [4,8,7,1]. In this paper the FSS idea is tested on the CPP problem.

T. Stützle, M. Birattari, and H.H. Hoos (Eds.): SLS 2007, LNCS 4638, pp. 212–216, 2007.

2   Packing Equal Circles in the Unit Circle

The problem of packing equal circles in the unit circle (PCC for short), introduced by Kravitz in [2], asks to position a given number of circular disks of equal
radius without any overlap within a unit circle, and to maximize this radius. Extensive bibliography, papers and test instances may be found for example on the
web at http://hydra.nat.uni-magdeburg.de/packing/cci/cci.html.

2.1   Mixed Coordinate Formulation

Let the set of disks to be packed be denoted by $I = \{1, \ldots, n\}$. A mixed formulation ϕ of the CPP problem is defined by splitting I into two (possibly empty)
parts C_ϕ and P_ϕ ($P_\phi = I \setminus C_\phi$) and to give each disk's center by its cartesian
coordinates when in C_ϕ and polar coordinates when in P_ϕ. Here the unit disk
is centered at the origin of both coordinate systems. Formulation ϕ is then the
following nonlinear program with $2n + 1$ real variables:

$$
\begin{cases}
\max r & \\
(x_i - x_j)^2 + (y_i - y_j)^2 - 4r^2 \geq 0 & \forall\, i, j \in C_\phi (i \leq j) \\
x_i^2 + y_i^2 \leq (1 - r)^2 & \forall\, i \in C_\phi \\
\rho_i^2 + \rho_j^2 - 4\rho_i \rho_j \cos(\alpha_i - \alpha_j) - 4r^2 \geq 0 & \forall\, i, j \in P_\phi (i \leq j) \\
\rho_i + r \leq 1 & \forall\, i \in P_\phi \\
(x_i - \rho_j \cos(\alpha_j))^2 + (y_i - \rho_j \sin(\alpha_j))^2 - 4r^2 \geq 0 & \forall\, i \in C_\phi, \forall\, j \in P_\phi \\
r \geq 0 & \\
x_i, y_i \in \mathbb{R} & \forall i \in C_\phi \\
\rho_i \geq 0, \qquad \alpha_i \in [0, 2\pi] & \forall\, i \in P_\phi
\end{cases}
\tag{1}
$$

   The first two constraints express that no two disks in C_ϕ may overlap, and that
all these disks should fully lie within the unit circle. The next two constraints do
the same in polar coordinates for disks in P_ϕ. The fifth set of constraints state
that no disk in C_ϕ may overlap with a disk in P_ϕ. Observe that the only linear
constraints are those in the fourth set. This shows that no two formulations are
linearly related.

2.2   Reduced Reformulation Descent

The choice $C_\phi = I$ defines the fully cartesian formulation ϕ_C, whereas $C_\phi = \emptyset$
defines the fully polar formulation ϕ_P. Reformulation descent, as introduced in
[6] uses only these two formulations of CPP. The local search for each formulation was a simple local minimization method of gradient type, in particular
we used Minos [3], a quite popular method of this type. Because these formulations did not allow us to tackle problems for large n, the number of constraints
being $O(n^2)$, we made some new experiments with reduced formulations. We sup-
pressed many of the non-overlap constraints by considering only such constraints
for pairs of disks not too far apart at the initial solution, more precisely when
their centers are at a distance $\leq 4r$. We found experimentally that the solution

quality remains the same, but since the number of constraints is considerably reduced, Minos is faster.

```
Function RFSS-PCC(n, kmin, kstep, kmax);
1  rcurr ← RD-PCC(n);
2  rmax ← rcurr; kcurr ← kmin;
3  let I be the set of all centers;
4  while Stopping Condition is not satisfied do
5      select subset P of kcurr centers at random; C = I \ P;
6      rnext ← MinosMixed(n, x, y, ρ, α, C, P);
7      repeat
8          rcurr ← rnext; P = C; C = I \ P;
9          rnext ← MinosMixed(n, x, y, ρ, α, C, P);
       until rnext ≤ rcurr ;
10      if rcurr > rmax then
11          rmax ← rcurr; kcurr ← kmin;
       else
12          kcurr ← kcurr + kstep;
13          if kcurr > kmax then
14              kcurr ← kmin;
           end
       end
   end
```

Algorithm 1. Reduced FSS for PCC problem

2.3   Reduced Formulation Space Search

Consider the set $\mathcal{F}$ of all mixed formulations. This corresponds to all choices of the index set C_ϕ, so has cardinality 2^n. $\mathcal{F}$ has a nested structure in $n+1$ levels, where each level is given by the cardinality of C_ϕ. For each formulation we use (reduced) RD as local search. The idea of FSS is that after each local search with end solution x, a new local search is started from the initial solution x, but using a new (reduced) formulation, randomly chosen from either level 1 if a new best result was found, or in the opposite case one (or $kstep$) level(s) up from the current level, until a maximum level $kmax$ is reached. This is more precisely described in the boxed pseudo-code Algorithm 1.

Illustrative example. We consider the case with $n = 50$. Our FSS starts with the RD solution illustrated in Figure 1, i.e., with $r_{curr} = 0.121858$. The values of k_{min} and k_{step} are set to 3 and the value of k_{max} is set to $n = 50$. We did not get improvement with $k_{curr} = 3, 6$ and 9. The next improvement was obtained for $k_{curr} = 12$. This means that a mixed formulation (??) with 12 polar and 38 Cartesian coordinates is used ($|C_\phi| = 38$, $|P_\phi| = 12$). Then we turn again to the formulation with 3 randomly chosen circle centers, which was unsuccessful, but obtained a better solution with 6, etc. After 11 improvements we ended up with a solution with radius $r_{max} = 0.125798$.

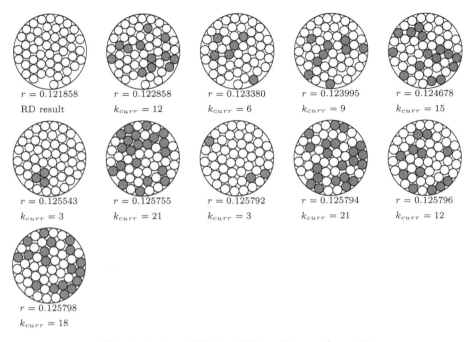

Fig. 1. Reduced FSS for PCC problem and $n = 50$

2.4 Computational Results

The FSS method was coded in Fortran and tested on a Pentium 3, 900 MHz-computer. Results in solving PCC problems by our Variable neighborhood FSS heuristic (Algorithm 6) are compared with the RD results recently published in [6]. They are presented in Table 1. In the first column the number of desired circles n is given, then the best known values from the literature for $1/r$. Columns 3 and 4 give the % deviations from these best known values for the best found and the average RD values, respectively, obtained in 40 runs of the code. Column 5 reports the corresponding average cpu time. The same values for FSS are given in the last three columns. It appears that the average error of the FSS heuristic is smaller, i.e., solutions obtained by FSS are more stable than those obtained with RD.

3 Future Research

A real variable neighborhood strategy [5] might be used in formulation space, by not working in levels around the fixed center ϕ_C, but rather allow re-centering around the previous formulation. Neighborhoods within formulation space are defined by way of the distance measure $d(\phi, \phi') = |C_\phi \triangle C_{\phi'}| = |P_\phi \triangle P_{\phi'}|$ where $\triangle$ denotes the symmetric difference operator between two sets.

Future research may also include other sets of formulations of CPP problems and use them within an FSS approach. For example, an unconstrained (min-max) formulations (with Cartesian and polar systems) may be used, then projective

Table 1. Packing in unit circle

		RD			FSS		
n	Best known	Best	Avg.	Time	Best	Avg.	Time
50	7.947515	0.06	0.79	3.19	0.00	0.24	80.54
55	8.211102	0.00	2.09	3.37	0.00	0.60	72.81
60	8.646220	0.03	1.40	4.71	0.00	0.95	84.39
65	9.017397	0.00	1.33	16.24	0.00	0.21	108.25
70	9.346660	0.10	0.99	19.56	0.01	0.27	151.64
75	9.678344	0.10	0.77	26.46	0.02	0.20	164.51
80	9.970588	0.10	0.93	39.15	0.04	0.23	229.49
85	10.163112	0.72	1.75	38.79	0.18	0.72	256.17
90	10.546069	0.02	1.27	96.82	0.02	0.56	294.77
95	10.840205	0.18	0.93	147.35	0.07	0.39	308.34
100	11.082528	0.30	1.01	180.32	0.12	0.68	326.67

(nonlinear) transformations among variables, etc. Instead of Minos, some other NLP solver may be tested. Extensions to more general circle packing problems with different radii might be considered, too.

References

1. Hertz, A., Plumettaz, M., Zufferey, N.: Variable space search for graph coloring, Les Cahiers du Gerad G-2006-81, Montreal, Canada
2. Kravitz, S.: Packing cylinders into cylindrical containers. Mathematics magazine 40, 65–70 (1967)
3. Minos, Stanford Business Software Inc. website, `http://www.sbsi-sol-optimize.com/products_minos5_5.html`
4. Mladenovic, N.: Formulation space search - a new approach to optimization (plenary talk). In: Vuleta, J. (ed.) Proceedings of XXXII SYMOPIS'05, Vrnjacka Banja, Serbia, p. 3 (2005)
5. Mladenović, N., Hansen, P.: Variable neighborhood search. Computers and Operations Research 24, 1097–1100 (1997)
6. Mladenović, N., Plastria, F., Urošević, D: Reformulation descent applied to circle packing problems. Computers & Operations Research 32, 2419–2434 (2005)
7. Mladenovic, N., Plastria, F., Urošević, D.: Stochastic formulation space search methods. In: Proceedings of EURO XXI - Iceland, pp. 79, Reykjavik (2006)
8. Plastria, F., Mladenović, N., Urošević, D: Variable neighborhood formulation space search for circle packing. In: 18th Mini Euro Conference VNS, Tenerife, Spain, November 2005 (2005)

Simple Metaheuristics Using the Simplex Algorithm for Non-linear Programming

João Pedro Pedroso

INESC - Porto and
DCC - Faculdade de Ciências, Universidade do Porto, Porto, Portugal
jpp@fc.up.pt

Abstract. In this paper we present an extension of the Nelder and Mead simplex algorithm for non-linear programming, which makes it suitable for both unconstrained and constrained optimisation.[1] We then explore several extensions of the method for escaping local optima, which make it a simple, yet powerful tool for optimisation of nonlinear functions with many local optima.

A strategy which proved to be extremely robust was random start local search, with a correct, though unusual, setup. Actually, for some of the benchmarks, this simple metaheuristic remained the most effective one. The idea is to use a very large simplex at the begin; the initial movements of this simplex are very large, and therefore act as a kind of filter, which naturally drives the search into good areas.

We propose two more mechanisms for escaping local optima, which, still being very simple to implement, provide better results for some difficult problems.

1 Extensions to the Simplex Method

Nelder and Mead's algorithm for non-linear programming [2] is a local search method for finding a minimum of a function, based on the movements of a simplex in a multi-dimensional space. These movements rely on function evaluations, and do not require information concerning the gradient of the function. Points of the simplex are ordered according to the value of the objective function to which possibly a penalty is added, if the problem is constrained and the solution is infeasible. However, the algorithm does not require the actual value of the function evaluation at each point; all that is required is to *order* the simplex's points, for determining through which vertex should reflection, expansion, or contraction be done. This is a common characteristic to all direct search methods [3].

The problem dealt with in this paper is characterised by a nonlinear function of a vector x, that we want to optimise. This function is usually multimodal, and in many cases non-smooth.

We assume that there are box constraints, i.e., for a problem of dimension N there will be constraints:

$$l_i \le x_i \le u_i \quad i = 1, \ldots, N. \tag{1}$$

---

[1] An extended version of this paper is available in [1].

T. Stützle, M. Birattari, and H.H. Hoos (Eds.): SLS 2007, LNCS 4638, pp. 217–221, 2007.

If in addition to the box constraints there are P more general constraints in the form $g_p(x) \leq 0$, for $p = 1, \ldots, P$, then the total constraint violation for a solution x can be assessed by

$$\delta(x) = \sum_{p=1}^{P} \max(g_p(x), 0) + \sum_{i=1}^{N} [\max(x_i - u_i, 0) + \max(l_i - x_i, 0)]. \qquad (2)$$

The comparison of solutions can be based on this value, as well as on the objective value. For two different solutions x and y, x improves y if and only if $\delta(x) < \delta(y)$, or $\delta(x) = \delta(y)$ and the objective value of x is better than that of y.

Based on this classification scheme, we are able to order the points of the simplex, even if some points are feasible and others not, and directly apply the simplex algorithm to constrained problems. Notice that equality constraints are poorly dealt by this modified method. The simplex will probably converge into a point which is on the surface defined by the equality, but will likely have much trouble for moving on that curve, in order to improve the value of the objective, without increasing infeasibilities. The classification system was used in [4]; a more elaborate method would be the filter method proposed in [5].

A random solution for a given instance is an N-dimensional vector that can be drawn as

$$x_i = \mathscr{U}(l_i, u_i) \quad i = 1, \ldots, N, \qquad (3)$$

where we denote a random number with uniform distribution in the interval $[a, b]$ as $\mathscr{U}(a, b)$. One possibility for using the Nelder and Mead algorithm is to start from a random point, as determined by Equation 3. Using the solution classification method described above, the search can start with a very large simplex, as this method tackles the possibility of getting out of bounds. Hence, we propose a setting were the initial step is very large: all the points except the initial random point will be out of bounds. Computational experiments have shown that this setup improves the overall performance of the simplex search; for smaller steps, the simplex would soon be trapped in a (generally poor) local optimum.

2   Escaping Local Optima

As the simplex method uses downhill movements, its solution will in general be a local optimum. If we wish to overcome this drawback, and be able to potentially obtain the global optimum of an NLP, we have to provide an escape mechanism.

The strategies that we describe for escaping are based on a restart criterion ϵ, and a stopping criterion M. Both of these are user-supplied values. Restart will occur if the vertices of the simplex have all evaluations which are feasible (or all infeasible), and the deviation between the objective (resp., the infeasibility) of the best and worst vertices is below ϵ. All methods will stop if the number of evaluations has reached the limit M.

2.1   Random-Start Iterated Simplex

This method consists of restarting the algorithm from a random solution every time there is convergence of the simplex according to the criterion ϵ, until the maximum number of evaluations M is reached. At each iteration, a random point is drawn and the simplex is reinitialised from that point (with a large step). Whenever the best found solution is improved, a new local search is performed with a smaller stopping criterion ϵ', for refining this local optimum.

The algorithm returns the best solution found across all iterations.

2.2   Directional Escape

Another possibility for escaping local optima is the following. When the simplex has converged according to the criterion ϵ, start expanding the simplex through its best vertex (always updating the ranking among the vertices). Expansion will initially decrease the quality of the point; but after a certain number or repetitions, we will reach a local *pessimum*, and the subsequent expansion will lead to an improvement. We propose to expand until the worst point of the simplex has been *improved*. At that point, we expect to be on the other side of the hill; hence, if we restart the simplex algorithm from that point, we expect to reach a different local optimum. We also restart if the bound has been crossed, using the first point outside bounds to initialise the simplex.

After an escape point is determined, the simplex is reinitialised around it by adding a large step independently to each of its coordinates. We called this strategy *escape*.

This strategy has the nice property of requiring no additional parameters.

2.3   Tabu Search

Tabu search for non-linear programming is not a precisely defined concept, as there is not a commonly used notion of *tabu* in this context. Actually, if the tabu concept is related to the kind of movement that is done, the escape mechanism described in the previous section can be considered as a tabu search: after a local optimum is reached, only expansion of the simplex is considered non-tabu.

In this section we propose a different concept: that of tabu based on the region of space being searched (as proposed also, for example, in [6]).

As we will shortly see, for tabu search to work in our setting it will have to be significantly modified, compared to the more usual tabu search in combinatorial optimisation.

Tabu solutions: classification. A trivial extension of the method devised for solution classification described in Section 1 consists of associating a tabu value to each solution, and then using this value as the primary key for solution sorting. In this context, for two different solutions x and y, x is said to improve y if and only if:

- x has a smaller tabu value than y;
- both have similar tabu values, and x has a smaller sum of constraint violations than y (i.e., $\delta(x) < \delta(y)$);
- both are feasible and not tabu, and the objective value of x is better than that of y.

Tabu regions. The most straightforward way of implementing a tabu search for continuous, non-linear problems is that of making the region around a local optimum (obtained by the Nelder and Mead algorithm) a tabu region. This way, for the tenure of the tabu status, we are sure that the search will not fall into the same local optimum.

This strategy, however, did not lead to good results, for the benchmarks used in this paper. We have tested many different approaches on this method, all of them with no success. The main reasons for this are related to the size of the tabu region: if it is too narrow, the search tends to find local optima on the border between tabu and non-tabu regions; on the other hand, if the region is too large, good local optima around the current solution are missed. This difficulty in parameterisation, and the observation that search around local optima is frequently essential to find better solutions, lead us to give up true tabu search, and try the opposite strategy: non-tabu search.

2.4   Inverting Tabu Regions: Non-tabu Search

As all the strategies that assigned a tabu status to the region of the last found local optima failed, we deduced that this tabu status barred the search from good regions, resulting in poor performance.

It is therefore expectable that for a good performance, the search has to be driven into the areas of previous local optima, instead of avoiding them. The rationale is that good local optima are often close to other local optima; hence, it might make sense to reinforce the search around previous optima, instead of avoiding regions close to them. Of course, the limit of this reasoning occurs when search cannot escape some particular local optimum.

In the algorithm that we devised for this, which could be named *non-tabu search*, the region around a local optimum is exploited by drawing a random solution on its vicinity, and restarting local search from it. A parameter σ controls the distance from the current base solution, used to draw new starting solutions. Another parameter, R, controls the number of attempts to do around each base solution. After R attempts are made, the base solution moves into the best solution found in theses tentatives.

These two parameters give a way for controlling the search, and to adapt it to the problem being tackled. Good parameters for a particular problem are generally easy to devise; but we could find no parameters that are simultaneously good for all the benchmarks.

3   Conclusions

In this paper we presented an extension of the simplex method for non-linear programming which allows its straightforward application to constrained problems.

For avoiding stagnation in local optima, we analysed the behaviour of several escaping mechanisms. The simplest of them is random-start local search. Another one was based on expanding the simplex from the local minimum, going uphill, until the expansion goes downhill again. At that point, we expect to be on the other side of the hill, and restarting simplex descent will likely lead to a different local optimum. The other possibility presented is based on the exploitation of the area of a the previous local optimum, by drawing starting points for local search in its vicinity: we called it non-tabu search.

Computational experiments (available in [1]) have shown that all the escaping mechanisms were effective for avoiding stagnation in local optima.

Due to the simplicity of its implementation, random start iterated local search is a highly attractive method. However, for some problems it is not able to find truly good solutions (though the average solution is generally of high quality).

The simplex expansion escaping mechanism is for most of the test cases slightly superior to random start local search, but in general the non-tabu search provides the best results.

Test and improvement of the escape methods for problems with equality constraints, and other possibilities of dealing with these constraints, remain as topics for future research. More research topics are their incorporation in more elaborate strategies, like strategic oscillation or population based methods.

References

1. Pedroso, J.P.: Simple meta-heuristics using the simplex algorithm for non-linear programming. Technical Report DCC-2007-06, DCC, FC, Universidade do Porto (2007)
2. Nelder, J.A., Mead, R.: A simplex method for function minimization. Computer Journal 7, 308–313 (1965)
3. Lewis, R.M., Torczon, V., Trosset, M.W.: Direct search methods: then and now. Journal of Computational and Applied Mathematics 124(1-2), 191–207 (2000)
4. Pedroso, J.P.: Meta-heuristics using the simplex algorithm for nonlinear programming. In: Proceedings of the 2001 International Symposium on Nonlinear Theory and its Applications, Miyagi, Japan, pp. 315–318 (2001)
5. Audet, C., Dennis Jr., J.: A pattern search filter method for nonlinear programming without derivatives. SIAM Journal on Optimization 14(4), 980–1010 (2004)
6. Chelouah, R., Siarry, P.: A hybrid method combining continuous tabu search and nelder-mead simplex algorithms for the global optimization of multiminima functions. European Journal of Operational Research 161, 636–654 (2005)

Author Index

Lecture Notes in Computer Science

For information about Vols. 1–4569

please contact your bookseller or Springer

Vol. 4616: A. Dress, Y. Xu, B. Zhu (Eds.), Combinatorial Optimization and Applications. XI, 390 pages. 2007.

Vol. 4615: R. de Lemos, C. Gacek, A. Romanovsky (Eds.), Architecting Dependable Systems IV. XIV, 435 pages. 2007.

Vol. 4613: F.P. Preparata, Q. Fang (Eds.), Frontiers in Algorithmics. XI, 348 pages. 2007.

Vol. 4612: I. Miguel, W. Ruml (Eds.), Abstraction, Reformulation, and Approximation. XI, 418 pages. 2007. (Sublibrary LNAI).

Vol. 4611: J. Indulska, J. Ma, L.T. Yang, T. Ungerer, J. Cao (Eds.), Ubiquitous Intelligence and Computing. XXIII, 1257 pages. 2007.

Vol. 4610: B. Xiao, L.T. Yang, J. Ma, C. Muller-Schloer, Y. Hua (Eds.), Autonomic and Trusted Computing. XVIII, 571 pages. 2007.

Vol. 4609: E. Ernst (Ed.), ECOOP 2007 – Object-Oriented Programming. XIII, 625 pages. 2007.

Vol. 4608: H.W. Schmidt, I. Crnkovic, G.T. Heineman, J.A. Stafford (Eds.), Component-Based Software Engineering. XII, 283 pages. 2007.

Vol. 4607: L. Baresi, P. Fraternali, G.-J. Houben (Eds.), Web Engineering. XVI, 576 pages. 2007.

Vol. 4606: A. Pras, M. van Sinderen (Eds.), Dependable and Adaptable Networks and Services. XIV, 149 pages. 2007.

Vol. 4605: D. Papadias, D. Zhang, G. Kollios (Eds.), Advances in Spatial and Temporal Databases. X, 479 pages. 2007.

Vol. 4604: U. Priss, S. Polovina, R. Hill (Eds.), Conceptual Structures: Knowledge Architectures for Smart Applications. XII, 514 pages. 2007. (Sublibrary LNAI).

Vol. 4603: F. Pfenning (Ed.), Automated Deduction – CADE-21. XII, 522 pages. 2007. (Sublibrary LNAI).

Vol. 4602: S. Barker, G.-J. Ahn (Eds.), Data and Applications Security XXI. X, 291 pages. 2007.

Vol. 4600: H. Comon-Lundh, C. Kirchner, H. Kirchner (Eds.), Rewriting, Computation and Proof. XVI, 273 pages. 2007.

Vol. 4599: S. Vassiliadis, M. Berekovic, T.D. Hämäläinen (Eds.), Embedded Computer Systems: Architectures, Modeling, and Simulation. XVIII, 466 pages. 2007.

Vol. 4598: G. Lin (Ed.), Computing and Combinatorics. XII, 570 pages. 2007.

Vol. 4597: P. Perner (Ed.), Advances in Data Mining. XI, 353 pages. 2007. (Sublibrary LNAI).

Vol. 4596: L. Arge, C. Cachin, T. Jurdziński, A. Tarlecki (Eds.), Automata, Languages and Programming. XVII, 953 pages. 2007.

Vol. 4595: D. Bošnački, S. Edelkamp (Eds.), Model Checking Software. X, 285 pages. 2007.

Vol. 4594: R. Bellazzi, A. Abu-Hanna, J. Hunter (Eds.), Artificial Intelligence in Medicine. XVI, 509 pages. 2007. (Sublibrary LNAI).

Vol. 4592: Z. Kedad, N. Lammari, E. Métais, F. Meziane, Y. Rezgui (Eds.), Natural Language Processing and Information Systems. XIV, 442 pages. 2007.

Vol. 4591: J. Davies, J. Gibbons (Eds.), Integrated Formal Methods. IX, 660 pages. 2007.

Vol. 4590: W. Damm, H. Hermanns (Eds.), Computer Aided Verification. XV, 562 pages. 2007.

Vol. 4589: J. Münch, P. Abrahamsson (Eds.), Product-Focused Software Process Improvement. XII, 414 pages. 2007.

Vol. 4588: T. Harju, J. Karhumäki, A. Lepistö (Eds.), Developments in Language Theory. XI, 423 pages. 2007.

Vol. 4587: R. Cooper, J. Kennedy (Eds.), Data Management. XIII, 259 pages. 2007.

Vol. 4586: J. Pieprzyk, H. Ghodosi, E. Dawson (Eds.), Information Security and Privacy. XIV, 476 pages. 2007.

Vol. 4585: M. Kryszkiewicz, J.F. Peters, H. Rybinski, A. Skowron (Eds.), Rough Sets and Intelligent Systems Paradigms. XIX, 836 pages. 2007. (Sublibrary LNAI).

Vol. 4584: N. Karssemeijer, B. Lelieveldt (Eds.), Information Processing in Medical Imaging. XX, 777 pages. 2007.

Vol. 4583: S.R. Della Rocca (Ed.), Typed Lambda Calculi and Applications. X, 397 pages. 2007.

Vol. 4582: J. Lopez, P. Samarati, J.L. Ferrer (Eds.), Public Key Infrastructure. XI, 375 pages. 2007.

Vol. 4581: A. Petrenko, A. Veanes, J. Tretmans, W. Grieskamp (Eds.), Testing of Software and Communicating Systems. XII, 379 pages. 2007.

Vol. 4580: B. Ma, K. Zhang (Eds.), Combinatorial Pattern Matching. XII, 366 pages. 2007.

Vol. 4579: B. M. Hämmerli, R. Sommer (Eds.), Detection of Intrusions and Malware, and Vulnerability Assessment. X, 251 pages. 2007.

Vol. 4578: F. Masulli, S. Mitra, G. Pasi (Eds.), Applications of Fuzzy Sets Theory. XVIII, 693 pages. 2007. (Sublibrary LNAI).

Vol. 4577: N. Sebe, Y. Liu, Y.-t. Zhuang, T.S. Huang (Eds.), Multimedia Content Analysis and Mining. XIII, 513 pages. 2007.

Vol. 4576: D. Leivant, R. de Queiroz (Eds.), Logic, Language, Information and Computation. X, 363 pages. 2007.

Vol. 4575: T. Takagi, T. Okamoto, E. Okamoto, T. Okamoto (Eds.), Pairing-Based Cryptography – Pairing 2007. XI, 408 pages. 2007.

Vol. 4574: J. Derrick, J. Vain (Eds.), Formal Techniques for Networked and Distributed Systems – FORTE 2007. XI, 375 pages. 2007.

Vol. 4573: M. Kauers, M. Kerber, R. Miner, W. Windsteiger (Eds.), Towards Mechanized Mathematical Assistants. XIII, 407 pages. 2007. (Sublibrary LNAI).

Vol. 4572: F. Stajano, C. Meadows, S. Capkun, T. Moore (Eds.), Security and Privacy in Ad-hoc and Sensor Networks. X, 247 pages. 2007.

Vol. 4571: P. Perner (Ed.), Machine Learning and Data Mining in Pattern Recognition. XIV, 913 pages. 2007. (Sublibrary LNAI).

Vol. 4570: H.G. Okuno, M. Ali (Eds.), New Trends in Applied Artificial Intelligence. XXI, 1194 pages. 2007. (Sublibrary LNAI).